AF443192

A NEW FORM OF WARFARE

The Rise of
Non-Lethal Weapons

A NEW FORM OF WARFARE

The Rise of Non-Lethal Weapons

Malcolm Dando

BRASSEY'S
London • Washington

First English edition 1996

UK editorial offices: Brassey's, 33 John Street, London WC1N 2AT
UK orders: Marston Book Services, PO Box 269, Abingdon OX14 4SD

North American orders: Brassey's Inc., PO Box 960, Herndon, VA 22070, USA

Malcolm Dando has asserted his moral right to be identified as author of this work

Library of Congress Cataloging in Publication Data
available

British Library Cataloguing in Publication Data
A catalogue record for this book is available from the British Library

ISBN 1 85753 127 2 Hardcover

Typeset by M Rules
Printed in Great Britain by Redwood Books Ltd, Trowbridge

CONTENTS

LIST OF TABLES

LIST OF FIGURES

PREFACE

There is widespread consensus in the Western strategic studies community that the state system remains in a transitional phase following the ending of the long East–West Cold War which has so dominated the second half of the 20th century. Despite the uncertainty about what form the state system will take in the future, there is also widespread agreement that for some time to come there are likely to be many more occasions when military forces from advanced industrialised countries are called upon to intervene in chaotic conflicts in other parts of the world.

As illustrated in Chapter 1, such interventions, exemplified by the peacekeeping operations in Bosnia and Somalia, have clearly been very difficult. Among the suggestions that have been put forward to help has been the unusual idea that operations could be more successful if they were rendered more benign through the use of novel non-lethal weapons. Any proposal to reduce the horrors inherent in such conflicts is worthy of careful consideration and, since the concept of non-lethality is developing towards the fielding of non-lethal weapons by the United States,[1] it is particularly necessary to subject it to thorough analysis. This is the aim of Chapter 2. Initially, a description is given of the new weaponry being considered, and then an attempt is made to discern the primary motive force behind this development. It is suggested that, rather than arguments for a more benign mode of peacekeeping being the driving force, the main reason for the rise of non-lethal weaponry may be the possibility of using it as an adjunct to regular military operations, as part of an effort to maintain military advantage through technological superiority. It is, in fact, just an aspect of high technology deterrence built on the revolution in military affairs.[2]

The problem with this approach to international politics is that other countries may choose to follow where advanced industrial countries lead. As the *Wall Street Journal* noted in 1993:[3]

> . . . the quiet US move into non-lethality could pry open a Pandora's box of chemical, biological and nuclear weaponry that diplomats have spent much of the 20th century trying to keep closed . . .

The idea that we in the West, far from being the champions of arms control, are at the root of the problem, is not well understood. This is illustrated in Chapter 3, which deals with the sorry history of attempts to constrain the use of inhumane weapons by the Inhumane Weapons Convention of 1980, and in Chapter 4, which recounts the consequences, in the years that followed, of the failure of the international community to regulate anti-personnel mines in that Convention.

As a report by the International Committee of the Red Cross pointed out in the run-up to the 1995 review of the Inhumane Weapons Convention:[4]

> The international community does not have to wait for catastrophes to occur, but can rather anticipate probable dangers. . . . Once a weapon is fielded it is very difficult to stem its proliferation and widespread use. Therefore it makes sense to dedicate some time to take preventive steps that would save enormous problems at a later stage.

Unfortunately, it does not appear that the widespread hopes of strengthening the Convention at the review, in order to deal with anti-personnel mines and the new danger of blinding lasers, have come to fruition. Unsatisfactory discussions on mines continued after the review and the new protocol agreed on lasers has an obvious loophole.[5]

Nevertheless it is important not to dismiss all non-lethal weapons, and the possibility of more benign peacekeeping operations, out of hand. What is needed is a critical analysis of each of the proposals in an attempt to discern the potential costs and benefits. With this in view, Chapters 5, 6, 7 and 8 deal with one of the most frequently mentioned non-lethal weapons – so-called 'incapacitating' chemical agents. One of the reasons for this choice is my own background in neuroscience and therefore particular interest in the subject, but more importantly, as the Chief Scientific Adviser to the UK Ministry of Defence acknowledged in 1993:[6]

> Two of the fastest moving areas of civil science are pharmacology and biotechnology . . .

What could emerge from Pandora's box in the way of weapons systems derived from these areas of research might therefore be particularly dangerous.

With the wealth of information publicly available in the open democratic society that is the United States, these central chapters of the book draw mainly on US material, but it is important to emphasise that similar developments are almost certainly under way in other countries, and the necessity of using this material should not be taken to imply that the United States is uniquely involved. Chapter 5 recounts the history of the use of incapacitant chemicals in warfare, paying particular attention to US experience prior to and during the Vietnam war. Chapter 6 gives a brief overview of the organisation of the human nervous system and Chapter 7 outlines our rapidly growing understanding of brain chemistry. Chapter 8 then provides an account of recent efforts by the US military to take advantage of this new scientific understanding to develop novel incapacitating chemical agents.

In Chapter 9 we return to the question of arms control and investigate how the attempt to preserve an 'incapacitating agents option' has left a large loophole in the Chemical Weapons Convention – the most effective potential arms control treaty that the international community has so far been able to agree. In this chapter a brief account is also given of how such multilateral agreements have to be improved in the future to really address the problems of regulating the worldwide spread of advanced industrialisation.[7] Chapter 10 returns to the general problem of the rise of non-lethal weapons and argues that the unthinking introduction of such weaponry into the armed forces of powerful states could indeed lead to a new form of warfare, but not necessarily the benign one we may anticipate.

With the wide range of subjects covered here, I could not have completed the book without help from a number of colleagues. In particular, I would like to thank Dr Julian Perry Robinson and his colleagues at the Science Policy Research Unit of the University of Sussex. I am not the first, and I shall certainly not be the last, to make use of the magnificent database on chemical and biological issues collected there by the Harvard Sussex Program on CBW Armaments and Arms Limitation. I should also like to thank Julian personally for guidance through the thickets of US arms procurement documentation. At Bradford, I have been lucky to work with Drs Nick Lewer and Steven Schofield on the issue of non-lethal weapons. Whilst their studies have a different focus from mine, the ability to share information and discuss interpretations has been most useful. Dr Steven Metz of the Strategic Studies Institute,

US Army War College, very helpfully introduced me to his and other US work on wider implications of this issue at an important point in the development of the book. I would also like to thank members of the NGO network for the help they have given with this academic study. My colleague Simon Whitby was able to provide me with crucial information while visiting the British American Security Information Council in Washington, DC during the summer and early autumn of 1995, and Simon Robinson and colleagues at Dfax kindly kept me aware of developments on non-lethal weapons and the review of the Inhumane Weapons Convention during 1995. I should, additionally, like to thank the staff of the J. B. Priestley Library at the University of Bradford where the Inter-Library Loans Office remained cheerful in the face of repeated protestations that this was the 'last pile' of requests. Lucy McKeever, the Life Sciences librarian, also organised the numerous database searches which generated much of the material covered in Chapter 8.

An earlier version of Chapter 2 has been published in *Brassey's Defence Yearbook 1996*, and an earlier version of Chapter 8 was presented, on the 50th anniversary of the United Nations, to a conference at Curtin University, Perth, Western Australia in May 1995.

Finally, closer to home, Janet Dando has turned my text into typescript, and readable English, and Owen Dando has designed and produced the numerous diagrams. I am very grateful to both for their help and good humour as I have tried to get a grip on the complex subject matter. Any errors that remain are mine alone. I should emphasise that I have tried to write for the general reader, and technical terminology has been kept to a minimum throughout. Numerous references and summary tables are provided as a guide to further investigation.

. 1 .

PEACEKEEPING IN CHAOTIC
CONFLICTS

Though in a less sorry state than their fellow peacekeepers who were chained to lamp-posts and bridges, there was no doubting, from the television images, that the soldiers of the Royal Welch Fusiliers were hostages of the Bosnian Serbs.[1] The United Nations operation in Bosnia, in late May 1995, was widely viewed as being in deep crisis and, for the first time since the Falklands War, the British Parliament was summoned specifically to 'consider the disposition of the British Army'. *The Guardian*'s political commentator, Hugo Young, summarised the resulting debate in desperate terms, calling every option a form of disaster.[2] His conclusion indicated the reason for the disaster with great clarity:

> . . .There is agreement on the ends, but also agreement that the means won't be available to achieve them. . .

Who could have imagined, in the euphoria which followed the end of the long East–West Cold War, that UN peacekeepers, so highly regarded that they were awarded the Nobel Peace Prize in 1988,[3] would so quickly reach this nadir?

The Expansion of UN Peacekeeping

United Nations peacekeeping, of a very carefully constrained kind, was largely successful during the Cold War but, in retrospect, it appears that in the early 1990s there was an over-estimation of what the world community might be able to achieve in the complex conflicts unleashed by the ending of that stand-off.

1

Traditional peacekeeping operations may have been complex to carry out but they were almost always quite clear in concept. Nation states which had grown weary of war would request help from the UN, and lightly-armed, impartial troops would be used to provide a buffer between the opposing forces.[4] The principles behind the deployment of UN forces were clearly understood to be:[5]

– consent of the parties to the dispute;
– non-use of force except in self-defence;
– voluntary contribution of contingents from small neutral countries;
– impartiality; and
– control by the Secretary-General.

During the 42 years from 1945 to 1987 some 13 UN missions were established around the world (Table 1.1), and the awarding of the Nobel Peace Prize of 1988 to the peacekeepers gives an indication of how these operations were viewed.

Table 1.1 UN peacekeeping missions initiated 1945-1987 [*]

Acronym	Area of Operation	Dates (established-terminated)
UNTSO	Middle East	1948–present
UNMOGIP	Jammu and Kashmir	1949–present
UNEF	Sinai	1956–1967
UNOGIL	Lebanon	1958
UNUC	Congo	1960–1964
UNSF	West Irian	1962–1963
UNYOM	Yemen	1963–1964
UNFICYP	Cyprus	1964–present
UNIPOM	India and Pakistan	1965–1966
DOMREP	Dominican Republic	1965–1966
UNEF II	Sinai	1973–1979
UNDOF	Golan Heights	1974–present
UNIFIL	Southern Lebanon	1978–present

* From reference 6

The first point to be grasped about operations since the end of the Cold War is the very large increase in their number.[5] While the exact figures are somewhat dependent on the criteria used, at least 22 operations were initiated between 1988 and October 1994 (Table 1.2). The increase in the scale of UN peacekeeping was emphasised in a special report by *International Defense Review*, which stated that:[6]

> . . . The UN has experienced a staggering increase in peacekeeping costs from US $800 million and 10,000 troops in 1990 to US $3.8 billion and over 70,000 troops in 1994 . . .

Given the long and successful experience in peacekeeping, and then the massive increase in available resources for the new operations, what has gone wrong? An answer to that question requires a more detailed examination of the nature of traditional peacekeeping and of the new operations and the context in which they have been carried out.

Table 1.2 UN peacekeeping missions initiated 1988–October 1994[*]

Acronym	Area of Operation	Dates (established-terminated)
UNGOMAP	Afghanistan and Pakistan	1988–1990
UNIIMOG	Iran–Iraq	1988–1991
UNTAG	Namibia and Angola	1989–1990
UNUCA	Central America	1989–1992
ONUVEH	Haiti	1990–1991
UNAVEM	Angola	1989–1991
UNAVEM II	Angola	1991–present
MINURSO	Western Sahara	1991–present
ONUSAL	El Salvador	1991–present
UNIKOM	Iraq–Kuwait	1991–present
UNAMIC/UNTAC	Cambodia	1991–1993
UNOMSA	South Africa	1992–1994
ONUMOZ	Mozambique	1992–present
UNOSOM	Somalia	1992–1993
UNPROFOR	Croatia, Bosnia and Macedonia	1992–present
UNOSOM II	Somalia	1993–present
UNOMIG	Georgia	1993–present
UNOMIL	Liberia	1993–present
UNOMUR	Uganda–Rwanda	1993–1994
UNMIH	Haiti	1993–present
UNAMIR	Rwanda	1993–present

* From reference 6

When the UN was founded, peacekeeping operations were not envisaged as part of its remit and they thus came about in an *ad hoc* way as urgent problems arose.[7] Their precise legal basis in the Charter of the UN is unclear. Chapter VI of the Charter is entitled 'Pacific Settlement of Disputes' and, under this heading, help could be given to parties who had decided that they wished to settle a dispute. Chapter VII of the Charter, entitled 'Action with Respect to Threats to the Peace, Breaches of the Peace, and Acts of Aggression', however, allows the Security Council to authorise distinctly more coercive action. Article 42 states that should sanctions etc. prove to be insufficient, the Security Council may:[8]

> . . . take such action by air, sea, or land forces as may be necessary to maintain or restore international peace and security. Such action may include demonstrations, blockade, and other operations by air, sea, or land forces of Members of the United Nations.

The use of armed forces in traditional peacekeeping operations was often said to take place under 'Chapter VI and a half', to denote the uncertainty but, given the fundamental consent of the parties, this was not a major problem. Moreover, the political stand-off between the two superpowers on the Security Council limited the number of possible operations.

The main reasons for the recent expansion in the number of peacekeeping operations has been the increasing ability of the Security Council to agree on taking action. Then, following the successful Chapter VII operation against Iraq in 1991, there was a fuelling of optimism about what peacekeeping might achieve. Adam Roberts noted:[7]

> . . . National governments, as well as the UN itself, have shared this mood to a surprising degree. The heads of government at the UN Security Council summit at the end of January 1992 and the Secretary General Boutros Boutros-Ghali in his *An Agenda for Peace*, published in June 1992, reflected and, for a period, reinforced this optimism.

And there were other successes, in Southern Africa for example, and the types of UN peacekeeping operation expanded to cover a wide range of problems (Table 1.3).

As more commitments were entered into, however, the severe limitations in the UN's ability to manage complex operations in the post-Cold War world were clearly exposed[9] and this brought the UN's standing into some disrepute. More seriously, intervention to deal with

Table 1.3   New types of UN peacekeeping*

– Monitoring and running elections.

– Protecting inhabitants of a region.

– Protecting 'safe areas'.

– Demilitarisation of areas.

– Guarding weapons.

– Assuring delivery of humanitarian aid.

– Assisting in reconstruction of governmental functions.

– Reporting violations of international law.

---

* From reference 7

the novel forms of chaotic *intra-state* conflict of the new era raised very difficult questions for an organisation based on 'the principle of the sovereign equality of all its members' (the words of Article 2.1 of the Charter) and whose Charter Article 2.7 states that:

> Nothing contained in the present Charter shall authorise the United Nations to intervene in matters which are essentially within the domestic jurisdiction of any state . . .

Recent peacekeeping operations, with their much wider aims, could therefore have been expected to be difficult, but to see just *how* difficult we need to examine the nature of the new kinds of conflict in more detail.

The New Conflicts

Of course, it is important not to dismiss the elements of continuity in peacekeeping operations. Not all deployments during the Cold War period were straightforward border-patrolling operations. The UN forces in Cyprus and Lebanon, for example, in part attempted to contain inter-communal violence in the hope that a political solution would be forthcoming.[10] Yet something new has been encountered in recent operations.

Helman and Ratner have argued that:[11]

> From Haiti in the Western Hemisphere to the remnants of Yugoslavia in Europe, from Somalia, Sudan, and Liberia in Africa to Cambodia in Southeast Asia, a disturbing new phenomenon is emerging: the failed nation-state utterly incapable of sustaining itself . . .

and they suggest that, as conditions deteriorate in such states, there will inevitably be increasing civil strife. There will inevitably also be calls from outside for 'something to be done'.

These authors suggested that we need to think of three categories of such failed states. Firstly, there are those like Bosnia, Cambodia and Somalia which have already failed – where the government has been overwhelmed by circumstances. Then there is a second group of states which are in the process of failing, such as Ethiopia and Zaire. Finally, there are some newly-independent states associated with the former Soviet Union whose fate is as yet unclear. Helman and Ratner suggested a variety of means by which the failure might be remedied but, characteristically, such failed states are divided societies with ethnic, tribal or factional groups which find it extremely difficult to agree acceptable means of living together. Divided societies certainly have a range of options available for achieving a solution to their divisions.[12] Rarely, however, does the stronger group make major concessions – as in Switzerland – in order to achieve consensus. Power-sharing and partition are difficult to arrange and repression by superior force, as in the former Soviet Union, merely contains the problem. Indeed, 'ethnic cleansing' has been resorted to in the past, as it has been recently in the former state of Yugoslavia, because at a certain level it can be seen to work.

This brings us to the crux of the problem: the characteristics of ethnic and factional conflicts. In the view of Cooper and Berdal,[13] such conflicts tend to be long-lasting and to involve insurgent movements as well as legitimate governments. Dealing with such movements can be very difficult, because obtaining agreement and consent to peacekeeping can be almost impossible and even if agreement is reached, it may well not endure. Then, for example, if peacekeepers:

> . . . find themselves in direct conflict with an ethnic insurgent group, it will almost always be an unequal struggle because the adversary will see itself as fighting for survival . . .

In such a situation the ethnic group will be prepared to sustain great losses, whereas the peacekeepers, in general, will not. Hence, as Hugo Young noted, there will be a gap between the ends that the international community wishes to achieve and the means it is willing to use. As became clear in Bosnia in 1995, if the international community decides that it should impose a solution rather than operate by consent:

> . . . the two are different operations requiring different levels of commitment, different military deployments and different rules of engagement.

For Cooper and Berdal, these new conflict situations therefore demand prior clear thinking about the political and military objectives of any peacekeeping intervention, and a deliberate matching of the intervention forces to the operational requirements. If this is not done disaster is the likely outcome. In the summer of 1995, the Bosnian Serb forces, despite intervention by NATO airpower, overran the UN-designated safe haven of Srebrenica and some 30,000 Bosnian Muslims were forced from their homes. UN spokesmen in Bosnia made it clear that they just did not have the necessary military forces available to carry out their mandate. States which agreed the mandate were unwilling to provide such troops,[14] and the whole UN operation had been dealt a terminal blow.

Will Interventions End?

Disasters in UN peacekeeping operations will not, of course, halt interventions by rich industrial states in other countries, as the intervention by NATO in Bosnia in late 1995 clearly demonstrated. When clear national interests are threatened – as oil supplies were when Iraq invaded Kuwait in 1990 – powerful states will find the means to achieve the desired ends. Moreover, there are other valid reasons for possible interventions, such as perceived threats to the international order or regional stability. But the reasons for interventions are also evolving. It is no longer considered legitimate to intervene in poorer states in order to collect debts, as it was in the last century, but humanitarian intervention is becoming more acceptable.[15] Indeed, as Mandelbaum has argued,[16] a reason:

> . . . for the shift in the international attitude toward intervention is the increasing acceptance of the protection of individual rights as an international norm . . .

That indeed conflicts with earlier ideas of inviolate state sovereignty, but it seems unlikely that the trend will be reversed and that people will be content to watch mayhem on their TV screens without wishing to help the unfortunate people caught up in terrible ethnic conflicts.

There will doubtless be more problems in poorer states facing economic difficulties, an escalating population explosion, resource depletion and environmental degradation, and continuing worldwide militarisation.[17] What is more, when the evidence is scrutinized, there is seen to be support for peacekeeping operations – even if difficult and dangerous – and even in the United States. Reviewing evidence from opinion polls about recent peacekeeping operations, Steven Kull concluded that:[18]

> . . . in most cases a majority of Americans favor US participation in UN peacekeeping operations. . . . a core of 45–50 percent consistently supports it . . .

Majority support is gained and increased if the mission is clearly perceived to be part of a UN operation, if participation is seen to be in the United States' interest, and if the operation is thought to have a chance of success. Majority support is practically certain if the President and Congress support participation. Furthermore, the evidence suggests that, contrary to popular opinion, US fatalities would not necessarily undermine that public support for peacekeeping.

Conclusion

It is possible that failures in peacekeeping, epitomised by events in Bosnia in mid-1995, will cause the international community to pull back from such operations. Greater restraint *is* probable in the near future but, having put such effort into increasing its peacekeeping activities, it seems very unlikely that the UN will abruptly cease such operations. The number of situations where peacekeeping will be called for is likely to increase rather than decrease and it seems most improbable that widespread public concern over human rights will decline.

It is certain that interventions around the world by powerful states will continue for other reasons such as concern for national interests or international stability. The dilemmas over how to intervene in an effective, but politically acceptable, manner are therefore not going to disappear overnight. The new world order will remain less than benign for many decades to come. The evidence available on rethinking within the UN[19] and in the United States[20] supports the view that, although caution has been induced by the failures of the first half of the 1990s, pragmatic multilateral action is still on the agenda.

.2.

BENIGN INTERVENTIONS WITH NON-LETHAL WEAPONS?

In view of the difficulties experienced by Western peacekeeping forces in the early 1990s, it is hardly surprising to find a variety of possible solutions being canvassed in the mass media. What concerns us here is the apparent rise of a new idea: benign interventions using non-lethal weapons. After reviewing the chaos of the Somalian intervention and the Waco siege, an article in *Newsweek*[1] in February 1994, with the title 'Soon, "Phasers on Stun"', asked the question:

> But what if US or UN forces – and what if the FBI – had an arsenal of tricky, new-tech weapons that could rout a mob, find and subdue hidden gunmen or fill an enemy fortress with a potent but harmless tranquilliser?

With the possibility of being able to keep the peace or fight without killing civilians, the article suggested, non-lethal weapons were attracting attention in the Pentagon and at the Department of Justice. Perhaps a little more cynically, *Harper's Magazine*[2] carried an article in October 1994 entitled 'Secret Weapons for the CNN (Worldwide Television) Era' in which it reported the view of Richard L. Garwin, an eminent US weapons expert, that non-lethal weapons could be:

> . . . extremely valuable in intervention where the 'CNN effect' puts a great premium on minimising or eliminating casualties, particularly among non-combatants.

We shall first review what these non-lethal weapons are, and then look in more detail at the causes of their increasing prominence, particularly in the United States.

The coverage of non-lethal weapons for benign interventions in the general media was preceded, earlier in the 1990s, by detailed articles in more specialist journals such as 'New weapons that win without killing on DoD's horizon' in *Defense Electronics*.[3] It has to be recognized though, as will later become apparent, that these technologies and ideas have a much longer history. On the military side, for example, in a 1968 article entitled 'Humane warfare for international peacekeeping', Lieutenant Colonel Celick noted that:[4]

> As far back as 1959, Major General William M. Creasy, a former Chief of the Army Chemical Corps, suggested that the development of psychochemicals provided a means of waging war without death . . .

From within the peace research community it was also being argued in the 1960s that:[5]

> . . . waging war without killing, mutilation, or destruction of property is technologically feasible. . . . The technological method suggested here for removing the undesirable features of war is Non-Lethal Violence . . .

The technologies suggested then to achieve this aim included:

> . . . sonic pistols firing invisible bludgeons of sound, tangle guns firing a mass of sticky entrapping threads . . . new hypnotic or psychological weapons.

What then of present possibilities?

Non-Lethal Weapon Technologies

Table 2.1 sets out a comprehensive range of the non-lethal weapon technologies discussed in the open literature during the early 1990s. A brief overview will be given of each of the technologies listed, to provide a factual basis for the discussion of non-lethal weapons and the implications of their possible use that follows. The boundaries of any classification of these weapon technologies are imprecise. To simplify the discussion somewhat, a broad classification, taken mainly from Alexander,[6] is used to group the various technologies (acoustic, electromagnetic etc.). It should be noted that there is obvious overlap with non-weapons aspects of standard military equipment (for example, electronic warfare systems or chemical weapon agent sensors) and standard police technology (for example, batons). For the most part, such non-lethal aspects of the use of force have not been included here. The focus is on those technologies which appear to be relatively new and which have been repeatedly mentioned in recent accounts of non-lethal weapons. A discussion of particular technologies and the implications of their use will necessarily involve a broader inspection of the employment of modern military forces and consideration of whether non-lethal means might require a wider reassessment of military operations.

Table 2.1 Non-lethal weapon technologies frequently mentioned in recent literature *

Category	Type		Description	Target
1. ACOUSTIC	1.	Infrasound	Low frequency, high intensity sound (modified to damage structures)	AP (AM)
	2.	Stun	More sophisticated system hitting target with physical force	AP
2. BIOLOGICAL/ MEDICAL	3.	Biodeterioration	Biological agents that degrade materials	AM
	4.	Incapacitating/ calmative substances	Chemicals that affect human behaviour	AP
3. CHEMICALS	5.	Adhesives	Chemicals that rapidly glue materials with strength	AM
	6.	Corrosives/ caustics	Chemicals that degrade materials	AM
	7.	Embrittling substances	Chemicals that reduce strength of materials	AM
	8.	Foams	Sticky and viscous or dense bubbles to immobilise people	AP
	9.	Lubricants	Chemicals that cause surfaces to become very slippery (anti-traction technology)	AM/AP
	10.	Vehicle engine modifiers	Chemicals that alter fuel combustion to stop engines	AM
4. ELECTRO- MAGNETIC	11.	Conductive materials	Long ribbons that short out electrical systems	AM

Table 2.1 *(Cont.)*

Category	Type	Description	Target
	12. Lasers		
	a. *Low energy system*	Hand-held laser rifle for temporarily blinding people	AP
	b. *Pulsed chemical laser*	A means of creating a high pressure shock wave in target hit	AM
	13. Light		
	a. *Munition*	Optical munition producing an omnidirectional flash of bright light	AP
		Optical munition producing a unidirectional flash of bright light	AP
	b. *Strobe*	Pulsed high intensity light to disorientate people	AP
	14. Microwaves	A single pulse of high power microwave energy (non-nuclear EMP)	AM
		High power microwaves repeated to form a pulsed beam	AM
	15. Stun	A variety of anti-personnel systems such as the 'taser' hand-held electrical stunner and a high-intensity 'dazzler' flashlight	AP
5. INFORMATION	16. Computer viruses	Introduction of viruses in order to cause computer systems to fail	AM
6. KINETIC	17. 'Bullets'	Physical objects which are designed to reduce the danger of injury whilst still hitting an individual	AP
	18. Entanglers	Nets, meshes etc. that can hold an individual (or vehicle)	AP (AM)

* Information derived from references in text

In November 1993 the US Army's Research Laboratory (ARL) held a 'Peacekeeping Technologies Seminar Wargame'. To prepare participants for the exercise a handbook was produced which described a wide range of non-lethal technologies. A copy of the handbook was obtained by the journal *Tactical Technology*, which enabled it to publish a more complete list of these technologies than is usually found in the general literature.[7] The technologies listed in the US Army handbook are the main source for the individual types set out in Table 2.1. In addition to regrouping the technologies into the broader categories suggested by Alexander, an attempt has been made to specify where the primary intent would be to attack *matériel* (AM) or personnel (AP). It must be understood, however, that one of the major criticisms levelled against non-lethal weapons is that AM weapons could have AP effects (and, in fact, be AMP weapons) and that AP weapons could, in some situations, have lethal effects. *Tactical Technology*'s original list gave over 30 possibilities,[7] but these have been reduced in Table 2.1 by concentrating only on the ones frequently mentioned in a selection of relatively detailed recent accounts.[1,3,6] – Where possible, an attempt has been made to simplify further by combining various similar types (eg lasers) under one heading. It will be obvious from Table 2.1 that even after these efforts at categorization and simplification, we are dealing with a widely disparate set of weapon system technologies. They differ, for example, in the complexity of the required technology, the amount of research and development that has been carried out, and the dangers inherent in their deployment.

The branch of science concerned with the study of sound is **acoustics** (Category 1, Table 2.1). Sound is produced by waves of energy passing through the air at a speed of about 300 metres per second. Normal human hearing detects wave frequencies between 20 and 20,000 cycles per second (in standard terminology, between 20 and 20,000 Hz where one cycle per second is termed 1 Hz). Ultrasonics refers to the study of frequencies above that range and infrasonics to frequencies below normal hearing (Figure 2.1).

As is well known from the trick of a soprano's voice breaking a glass, if sound bears the correct frequency relationship to an object it can set up a resonance which makes the structure vibrate and fracture. This is the basis of the idea that, if a strong enough signal could be produced, then a means would be available to damage built structures. It is already possible to use a suitable tuned beam of low-frequency, high-intensity sound to affect people with extreme nausea by setting up vibrations in the organs of the inner ear. Reports suggest that Scientific

Figure 2.1   The Electromagnetic Spectrum

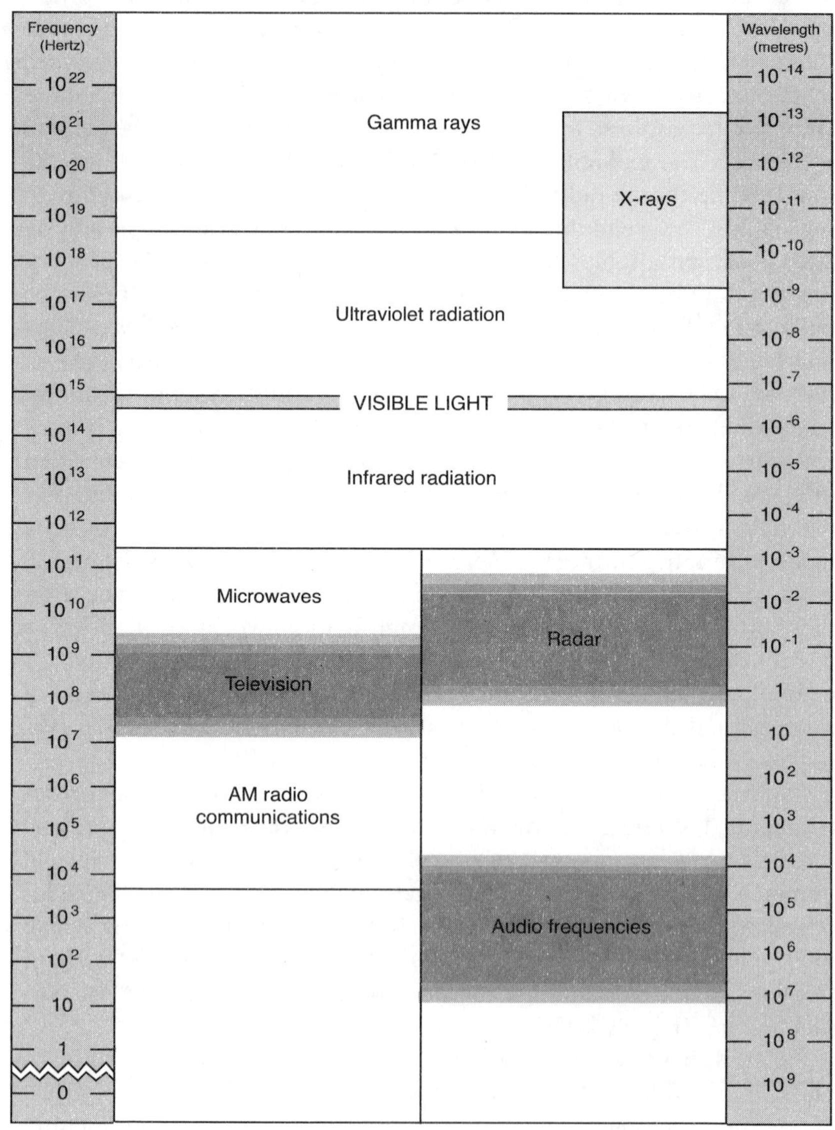

Applications and Research Associates (SARA) of Huntington Beach, California, in association with Los Alamos National Laboratory (LANL) and the US Army's Armament Research, Development and Engineering Center (ARDEC) are involved in developing such a beam weapon.[3] The equipment required to attack human beings would be quite large and complex,[9] but the idea is that it could be used, for example, to keep angry crowds like those in the 1979 assault in Tehran away from a US Embassy. More sophisticated systems in which the impact of the sound waves was enhanced, for example by resonance between waves from different sources intersecting with each other, are also possible and are being investigated.

While it is argued by supporters[14] that the effects on humans cease when the beam is turned off 'with no lingering physical . . . damage', the *SIPRI Yearbook*[15] noted that 'they can, however, inflict serious damage in internal organs at short range'. This problem, of the potentially uncontrollable effects of weapons which may well be intended to cause little lasting harm, is a recurrent theme in the consideration of non-lethal weaponry. It relates rather directly, of course, to the requirement in international humanitarian law of war that weapons should not be indiscriminate in their effects.

It is frequently reported that **biological** agents (Category 2, Table 2.1) are being considered as non-lethal weapons. The *Bulletin of the Atomic Scientists*[16] noted that LANL had carried out a review of potential organisms, and quoted from a publication produced by the laboratory; 'As a result, we discovered a bacterium that degrades a specific material used in many weapon systems'. While it is certainly true that industry is making use of bacteria for a wide variety of purposes, and that this use will surely increase as genetic engineering capabilities develop, even proponents of non-lethal weapons[6] acknowledge that this particular use of biological agents is 'politically the most sensitive area'.

Supporters of efforts to restrict worldwide militarisation through the development of international arms control regimes are currently hoping that the potential impact of the new biotechnology revolution (genetic engineering) can be reduced by the addition of an effective verification protocol to the Biological Weapons Convention (BWC). Use of bacteria for non-lethal weapons poses a direct threat to the BWC. Article 1 of the convention clearly states:

Each State Party to this Convention undertakes never in any circumstances to develop, produce, stockpile or otherwise acquire or retain:

1. Microbial or other biological agents, or toxins whatever their origin or method of production, of types and in quantities that have no justification for prophylactic, protective or other peaceful purposes.

To use bacteria in non-lethal weapons to degrade critical weapon components or to destroy fuel supplies[13] would therefore appear to require the United States to scrap the basic condition of the BWC – not quite the best way to convince a suspicious world that it actually favours making the Convention an effective instrument of arms control!

Consideration of the possible consequences of using **chemicals** to alter human behaviour (Category 2, Type 4, Table 2.1) will form a major part of Chapters 5–8 of this book. We will only note here that so-called 'calmative' agents to induce sedation, absorbed quickly via the skin through admixture with dimethyl sulphoxide (DMSO), are frequently mentioned along with other incapacitants and irritants as potential non-lethal agents in the current literature.[7-11,13-16] As will be argued in detail later, these agents pose grave problems in regard to possible indiscriminate effects and, in the guise of 'advanced riot control agents for law enforcement purposes', could be extremely damaging to the effective entry into force of the Chemical Weapons Convention (CWC)[16] on which so much international effort has been expended, successfully thus far, since the end of the Cold War. To allow an unreflective dash for preparation to use such chemicals, which would destroy what could be a cornerstone international agreement in the construction of a stable new world order over the next few decades, would appear to be perverse in the extreme.

A variety of other chemicals have been proposed for non-lethal weaponry (Category 3, Table 2.1). One idea is to use modern adhesives to glue military vehicles in place,[7,11] or to foul the turbine blades of hydroelectric power stations,[13] or sprayed as a mist from an aircraft to foul the engines of enemy planes.[9] Whilst the primary intention appears to be to attack *matérial*, sticking people is certainly also envisaged. As Paul Evancoe, lately Deputy Director for Special Operations at the US State Department's Office of the Coordinator for Counter-Terrorism, commented in 1993:[9]

> . . . Against people, polymer agents could be employed to glue a person to almost anything that he or she may touch, including another person.

A moment's reflection, or the experience of having inadvertently misused a tube of superglue, would surely raise in most reasonable minds difficult questions of whether such systems could be used with any discrimination in some of the proposed applications.

These questions must surely arise in a particularly acute form in regard to the use of extremely corrosive or caustic chemicals (Category 3, Type 6). The terminology applied can be confusing, but the term 'supercaustics' appears to be used to cover both supercorrosives (acids) and supercaustics (bases).[13] By mixing hydrochloric and nitric acids, for example, a reagent is formed which will dissolve both metals and organic substances such as plastics or rubber, while the bases sodium hydroxide, potassium hydroxide and caesium hydroxide could be used, for example, to attack glass in optical systems. To anyone who has handled such chemicals, as most of us have at school, the proposed uses beggar belief. Evancoe states:[9]

> These powerful agents could be formulated as liquids, sprays, powders, or gels to be dropped by aircraft, dispersed from an artillery round or applied by soldiers . . .

International humanitarian law of war stipulates that weapons should not be indiscriminate, neither should they cause unnecessary suffering. How civilians might be protected from the possible consequences of an aerial attack with a supercorrosive mixture of hydrochloric and nitric acids is a question that does not appear to admit a sensible answer.

So-called embrittlement agents are generally referred to as Liquid Metal Embrittlement (LME) agents and are metals (or alloys) which are liquid at near-normal temperatures and reactive with other metals. Examples are mercury, caesium, gallium and rubidium. The process envisaged is as follows:[13]

> . . . The amalgam-like alloys that are formed as the liquid is absorbed [by the material of the attacked structure] can cause the material to be weaker and/or much more brittle . . . under loading, a catastrophic failure can result.

It is suggested that the agents would be very difficult to detect once applied (say in a covert operation) and that different agent/structural material combinations would allow the process to occur quickly or relatively slowly. While painting LMEs on a bridge would appear to be a relatively safe, discriminating application,[8] Evancoe's views suggest a much more indiscriminate approach:

> . . . When coupled with liquid metal embrittlement technology C+ [supercaustic] has tremendous potential across a wide spectrum of target applications such as weapons, vehicles, buildings and equipment.

The listing in *Tactical Technology*[7] suggesting delivery by 'hand, mortar, artillery' would also support the view that potentially indiscriminate applications are envisaged. Thus similar reservations to those advanced in regard to adhesives and corrosives/caustics would be in order.

It appears that in its efforts to ensure protection of convoys carrying nuclear materials (where it is dangerous to use firearms) the US Sandia National Laboratory developed two types of foam. One was very sticky and viscous and the other a very dense bubble foam. These could be used if a transport container were subject to terrorist take-over and would, it was envisaged, imprison the attackers so that they could do no harm.[13] The sticky foam is reported to be under development for squirting at rioters or suspects fleeing from the police. However, the question of what should then be done with a rioter, suspect or (potential) enemy soldier who is covered in sticky glue does not seem to have been adequately addressed.[10]

In what is termed anti-traction technology, the large-scale use of a variety of new lubricants is being considered for interference with military operations. Clearly, if a runway, railway or road system could not be used because the surface was too slippery, logistical support of forces would be rendered much more difficult. Supporters of this kind of non-lethal weapon technology[14] refer to 'environmentally neutral lubricants' but also point out that such materials would prevent operations for some considerable time because they are 'costly and time-consuming to remove'. According to the information obtained by *Tactical Technology*,[7] applications would be by 'mortar, artillery, vehicle, aircraft, missile', and possibly on a large scale.[9] The nearest analogy in regard to cleaning up afterwards is probably the treatment of a large-scale oil slick on land. It is therefore at least possible that major environmental problems could arise, if not in relation to the original attack, then at least with respect to the problems left behind for local inhabitants.

Modern forces are extremely dependent on vehicles powered by internal combustion engines. It is well known that a variety of chemicals can be used to alter the combustion characteristics of fuels so that engines cease to function. It has been envisaged that gases could be put into mines which would then 'non-lethally' stop a convoy,[14] but the hilarious account by the BBC TV defence correspondent, David Shukman, of a meeting with the inventor of one such non-lethal weapon suggests that considerable development work may still be required.[17] Other means of interrupting the operation of engines, such as chemical viscosification[8] or bacteriological degradation of fuel[13] or materials that would clog air-filters[8] have also been suggested.

Since the predominant technological revolution affecting military operations at present is based on electronics,[17] it is no surprise that many of the suggested non-lethal weapons in the literature come under Category 4 in Table 2.1 (**electromagnetic**). The Electrical Power Distribution Munition (EPDM) was used in the first days of the 1991 Gulf War.[13] Long, conductive, carbon ribbons were dropped onto Iraqi power lines and distribution points, causing massive short-circuiting. Even when the material was cleared from the lines, it required only a breeze to lift ribbons from the surroundings and cause further short circuits. Unfortunately for advocates of non-lethal weaponry, US generals were unsure about the effectiveness of the weapon and so the same targets were also attacked directly with conventionally (explosive)-armed cruise missiles within 24 hours.[18] There are reports in the literature[8] of the potential use of conductive particles as well as ribbons. While the ribbons would cause problems for those attempting to clear up afterwards, the spreading of large quantities of small conductive particles might also constitute a biological (respiratory) hazard.

It has been reported that some advocates of non-lethal weapons wanted the US President to declare a Non-Lethal Defense Initiative, similar to the Strategic Defense Initiative (SDI), in order to give proper focus and backing to this defence expenditure. SDI certainly gave impetus to the development of laser systems and the literature on non-lethal weapon technologies is also dominated by discussion of lasers. Two main types of weapon are clearly envisaged, as indicated in the subdivisions of Type 12 in Table 2.1.

Ordinary light radiates in all directions at many different wavelengths. Laser light is very different. A laser is a device which produces an intense beam of parallel light with a defined wavelength. The name is derived from 'light amplification by stimulated emission of radiation' (laser). The special properties of laser light have been used in many applications. Since a laser beam spreads very little as it travels, it can be focused to a small intense spot and used, for example, to refix a detached retina in the human eye by a process akin to spot welding.

Laser weapons designed to dazzle the pilots of enemy aircraft have been deployed for some years by the British Navy.[13] However, in many ways the most obviously worrying non-lethal weapon possibility is the hand-held laser. As will be explained in Chapter 3, the potential large-scale deployment of such weapons, which can undoubtedly be used deliberately or inadvertently to blind people, is of great concern to the International Committee of the Red Cross. *Defense Electronics* reported in early 1993 that laser rifles were being developed by the McDonnell

Douglas Company in an outgrowth of an older US Army research programme called COBRA, designed to provide a system for blinding enemy optical sensors. The production of a laser rifle had become possible a few years before the *Defense Electronics* article and the whole research programme, parts of which had previously been unclassified, were reclassified in the so-called 'black' or secret part of the defence budget.[3]

More recent reports[9,13] concern a Low Energy Laser (LEL) system that has the capability to oscillate its output through red, yellow, green, blue and ultraviolet light. The countermeasure to a laser which produces a particular colour of light is to wear a pair of goggles which filter out that colour. As Evancoe notes:[9]

> . . . because the LEL color range oscillates, a terrorist's eyes would have to be protected by filters (goggles) covering this broad spectrum. Such filtering lenses are expensive, bulky to wear, and available only from a sophisticated manufacturer.

Of course, even if the laser were set only to blind a victim temporarily, should that person be using a night vision device 'all bets are off', according to one inside source.[3] Such night vision devices greatly magnify available light and are widely used by military forces, so inadvertent blinding would be a real possibility. Yet it appears that such a system was tested by the US Army.[3] According to the 1994 *SIPRI Yearbook*, a contract dated 1992 planned for $80 million to be spent by 1998 on a hand-held laser for damaging sensors and human eyes. The necessary technology has been made available by advances in solid-state laser research.[15] The second laser weapon frequently referred to in the recent literature is a pulsed chemical laser. In this system, being investigated by the US Army's Armament Research, Development and Engineering Center and Los Alamos National Laboratory, the intention is to generate a very-high-power pulse of energy which produces a shock wave in the target.[13] It appears that the energy levels required are feasible, and it is suggested that controlled effects can be achieved on *matériel* and people.[3] SIPRI stated that trials on a compact deuterium fluoride prototype laser were carried out in 1993.[15] Currently, a fuller account of the numerous US laser programmes is being developed.[18]

Effects somewhat similar to those of lasers can be obtained with so-called optical munitions. *Tactical Technology* reported in 1993 that exploratory development of both an isotropic (omnidirectional) and a unidirectional radiator system was on-going.[8] The omnidirectional radiator consists of a gas which is compressed at a very high temperature so that it becomes incandescent. The flash produced is very bright and across a very broad spectrum of light. In the directional version a laser dye rod is incorporated into the munition and the light is emitted in a single direction. Evancoe noted:[9]

> . . . These special munitions, that are designed to radiate directionally or omnidirectionally already exist and are fired from conventional weapons. This technology can be packaged for air drops, fired aloft from mortars or artillery . . .

He also suggested that such weapons may have been used by Soviet forces in Afghanistan, where 'Doctors treated scores of Afghani villagers afflicted with various degrees of blindness characteristic of laser or isotropic radiation retinal damage following Soviet offensive operations', and expressed concern over possible proliferation of such weapons as the former Soviet Union disintegrates. If Evancoe is correct in his suggestion that such munitions used in Soviet operations affected Afghan villagers in this way, it seems possible that similar problems may arise if they are used in US operations in the future. Moreover, while such weapons could conceivably be used in some kind of grenade in a hostage release attempt, the references to delivery by artillery or by air surely suggest that the US Army is thinking more in terms of usage as an adjunct to standard military operations (what has been called 'pre-lethal usage'[16]) rather than as some entirely new doctrine of non-lethality. This issue will be explored further at the end of the chapter.

A perhaps more commonplace use of light is in strobe lights. The intent here is to make use of the Bucha effect, whereby high intensity lights flashing at or near brain-wave frequency cause vertigo, nausea and disorientation. The problem is that this stimulus may trigger epileptic seizures in a percentage of the population. Other possible non-lethal weaponry discussed, involving light and vision, are holograms to project images, and various forms of active camouflage such as paint which changes colour and pattern depending on lighting and temperature.[7,9,11,13]

Delivery of high frequency microwaves (Figure 2.1) to unprotected functioning electronic systems damages these systems severely, a fact known from work on the Electromagnetic Pulse (EMP) derived from nuclear explosions. Reports abound of efforts by advanced industrial nations to create such an EMP by non-nuclear means.[7,9,13,15,19] According to the 1994 *SIPRI Yearbook*, for example:

. . . increasingly sophisticated technology in advanced industrial nations, notably the USA, the UK and Russia, is now enabling an EMP to be produced from conventional explosives in a relatively small warhead.

HPM [High Power Microwave] warheads have been under development in classified programmes in the USA for a number of years . . .

The reports suggest that the warhead is small enough to be fitted to a cruise missile. Details of the programme were given in an article in *Aviation Week and Space Technology* in 1993.[19] Work on both civil and military applications of non-nuclear EMP generators is being carried out at Los Alamos, according to the article. Scientists describe a system in which a coil is wrapped around a copper cylinder full of explosives. Capacitors supply an initial current to the coil and this creates a magnetic field which is then compressed by the explosion to deliver a short-duration pulse output of great power. The military version is also said to incorporate explosives outside the coil to increase compression and thus directional output of the EMP pulse. Also important is a well-tuned antenna which focuses the output within about a 30-degree arc over a range of several hundred metres. There are also, reportedly, efforts to produce a system which could deliver repeated pulses of high-energy microwaves[14], but this would appear to require very large static installations[13] or considerable technological advances.[9] It should be noted that, according to SIPRI, 'These weapons can also cause unconsciousness . . . by upsetting the neural pathways in the brain'. A variety of electrical devices are in use or under consideration for stunning or disorientating individuals. These include the taser hand-held electrical stunner and a very-high-power battery-powered flashlight (the dazzler).[14]

We have specifically excluded the wide range of Electronic Warfare (EW) systems available to military forces as not really appropriate for consideration under the label of non-lethal weaponry. However, frequent reference is made in the available open literature to the use of **information** (Category 5, Table 2.1) such as computer viruses and hidden codes which can be activated in enemy systems if necessary.[7,13,14,16] Some sources also refer to the use of advanced electronics for the takeover of enemy communication systems and then the selective broadcasting of information chosen by the intervention forces.[17] It is also advocated by some people[6] that 'There should be a strong tie to psychological operations' in non-lethal operations.

A final category of non-lethal weapon system is termed **kinetic** (Category 6, Table 2.1). Alexander[6] defines this as 'the use of high-strength materials to stop machines functioning'. I have widened the definition here to include the use of moving objects or entanglements for use against people. The generic problems with all such technologies are well known from the wide experience that police forces have had with rubber bullets. The weapon has to have a certain level of power in order to be of any use, so there is always the danger of incorrect use (a rubber bullet fired at too close a range), or of the non-lethal weapon producing a fatality in a very susceptible individual. Clearly, problems with rubber bullets have led to efforts to develop less dangerous systems (so-called 'bean bags') in which there is a reduced chance of injury. Various forms of net have been devised to restrict the movement of people or vehicles,[13] and simple methods of fouling propellor blades proposed.[14]

From weapon systems which appear more suited to strategic first strike attacks at one end of the range through to covert operations at the other, it is difficult to argue that what is being discussed results from a coherent policy on non-lethality related to potentially more benign interventions in the future. In order to understand what is behind the upsurge of interest in non-lethal weapons technology in the United States in the early 1990s, we therefore have to enquire in more detail about the people and organisations involved.

The 'Boosters' of Non-Lethality

The two people most frequently mentioned as promoters of the idea of non-lethal weapons, and of interventions by US and allied forces using these weapons, are Janet and Christopher Morris of the Washington-based US Global Strategy Council. There can be no doubt that they have a broad vision of the dangerous world that the ending of the Cold War has produced and of the need for a new approach to the inevitable conflicts that will occur.

As set out in their 1993 paper with Victor Krivorotov, 'The age of chaos: threat and solution beyond containment', there are many points with which an idealist hoping for a more just and peaceful society might agree.[20] It is argued that the ending of the Cold War lifted constraints from long-running conflicts with deep roots, in both the Third World and the former Soviet Union. Poverty and militarisation will exacerbate these conflicts, and abuses of human rights are inevitable. Moreover, the industrialised nations, armed to the teeth with weapons suitable for an East–West war, will often be reluctant to carry out the kind of intervention undertaken against Iraq in 1991, both because the level of casualties would become increasingly unacceptable and because such interventions would often be destabilising. The authors argued:

> . . . The United Nations, the United States, and the international com-
> munity can combat the growing threat of global chaotic destabilisation
> by instituting a doctrine of non-lethal force and implementing a pol-
> icy of nonlethality . . .

They further argued that world citizenship is even within our grasp if this approach is taken. The suggested approach is certainly comprehensive:

> . . . This policy must include strategic doctrine mandating the devel-
> opment and maintenance of all appropriate technological capabilities
> necessary to overcome the surge in violent crimes against humanity
> following the dissolution of the bipolar balance of power . . .

Thus the wide range of non-lethal technologies previously discussed could fit within this expanded vision of peacekeeping – which might involve strategic action, it appears, so long as it was non-lethal. A proposal was also included for the gradual conversion of military industry by an initial switching to non-lethal projects.

Leaving aside the question of whether *any* solution to the problems of the poorer parts of the world is possible without a fundamental redistribution of wealth and resource usage,[21] the Morrises' prescription becomes a lot less benign (non-lethal) when they get down to details. As they explained in a 1994 publication:[14]

> . . . there should never be a 'non-lethal' unit fielded by the US. . . .
> Rather, non-lethal weapons must be capable of precise non-lethal-to-
> lethal responses, so the enemy will not know what response to expect,
> and so US field commanders can have a full panoply of responses
> available.

Perhaps this less than non-lethal approach was necessary as a means of getting their ideas accepted in Washington policy-making circles, an enterprise at which the Morris team appears to have been extremely successful.

The Global Strategy Council is chaired by Dr Ray Cline, former Deputy Director of the CIA, and well connected in Washington.[17] In part because of lobbying by this conservative think-tank, President Bush's Defense Secretary, Dick Cheney, set up a Non-Lethal Warfare Study Group under the direction of Paul Wolfowitz, his Under-Secretary for Policy.[22] Although supporters argued for a high-level, co-ordinated Non-lethality Initiative, it appears that this study did not reach fruition because of a variety of disagreements within the Pentagon.[3,10,15,17] Nevertheless, after the new Clinton administration took office it seems that there was a decision to increase both efforts in this area and the secrecy about what was being undertaken. The Morris team has objected to this approach, suggesting that it was taken to protect the budgets of national laboratories such as Los Alamos from Congressional scrutiny. However, another prominent supporter of non-lethal weapons, John Alexander, formerly Disabling (Non-lethal) Technologies Program director at Los Alamos, has argued that such secrecy is necessary in order to prevent potential opponents developing countermeasures.[10] Alexander, who has a military and special operations background, clearly sees non-lethal weapons as providing extra options for military commanders across the complete range of conflicts 'from operations-short-of-war to high-intensity conflict', and having diverse functions 'from peacekeeping capabilities to strategic paralysis of an adversarial nation-state'. This approach begins to make sense of the very wide range of technologies discussed under the heading of non-lethality. We are *not* dealing with ideas of better, more benign, peacekeeping but with an expanded view of standard military operations. We need therefore to look more closely at developments within the US military.

It is well known that in the early stages of the debate the US Army's Training and Doctrine Command (TRADOC) issued for comment a draft document[23] entitled 'Operations concept for non-lethal capabilities'. More significant, perhaps, is the investment in non-lethal technologies by the Army's Armament Research, Development and Engineering Center in its Low Collateral Damage Munitions (LCDM) programme. In this[3] there is said to be research into new technologies which may be weaponised to 'effectively disable, dazzle or incapacitate aircraft, missiles, armoured vehicles, personnel and other equipment while minimising collateral damage'. Significantly, the LCDM projects come within a Joint Services Small Arms Program which seeks capabilities from disabling to lethal, indicating, as noted in detailed reviews, that these non-lethal weapons will probably be regarded as an adjunct to regular military lethal weaponry.[13]

Additionally, there is evidence of major links in big programmes between the Army (and Air Force) and the National Laboratories at Los Alamos and Lawrence Livermore, and there is considerable evidence of industry interest in non-lethal weapon technologies.[24] Meanwhile, studies of non-lethal technologies have continued within the administration. In April 1994 a five-year agreement was signed by the Department of Defense and the Department of Justice to address common problems. These included:[12]

Mission kill/less-than-lethal technologies to expand the spectrum between the gun and personal persuasion.

The overall impression therefore has to be that, behind a veil of considerable secrecy,[16] a major effort is under way which is likely to produce a range of new weapon systems in the coming years. We should now step back from the immediate issues to examine the general context within which these technological advances are occurring and to consider the possible implications.

General Context and Implications

During the prolonged crisis in Bosnia some public consideration was given to the possible use of non-lethal weapons by US forces, should they become directly involved. In August 1992, for example, *Aviation Week and Space Technology* carried a report[25] entitled 'US weighs use of non-lethal weapons in Serbia if UN decides to fight'. This article appears to have been prompted by statements from Senator Sam Nunn, at that time Chairman of the Senate Armed Services Committee. Carbon-fibre warheads and EMP warheads on cruise missiles to shut down Serbian electrical supplies and disrupt local command and control were discussed. Additionally described were chemicals sprayed on runways to destroy aircraft tyres and microbes for turning fuel supplies to useless jelly.

What was probably a more realistic scenario was put forward in November 1992 in an article in the *New York Times* co-authored by M. J. Dugan, a former US Air Force chief of staff.[26] The article, perhaps significantly titled 'Operation Balkan Storm: Here's a plan', envisaged 'a three-step plan based on the use of American competitive advantages'. The final step of the plan was 'active belligerency, in two phases: first, destroying Serbian forces in Bosnia and, second, using concentrated force against Serbia itself'. In the first phase US planes would establish air supremacy and allied forces would smash Serbian forces from the air – all by conventional, lethal means. In the second phase a mix of conventional means and non-lethal weapons would be used. According to the article: 'Technology using carbon-fiber strands allows us to render useless Serbian's electricity grid'; 'Other technology allows us to turn petroleum products in refineries and storage tanks to useless jelly'; and 'we take over Serbian air waves to make our operations to end the war clear to the Serbian people'. This overall context, of using the technological edge of the US and other industrial allies in future warfare, is crucial to understanding the real meaning of non-lethal technologies. We should briefly consider this more general issue before discussing the possible implications of current efforts to pursue that part of *competitive technological advantage* devoted to weapon systems which may (for whatever reason) be labelled as non-lethal.

We may gain an insight into high-level Western thinking on this issue from an article written by Professor Sir Ronald Oxburgh, Chief Scientific Adviser to the UK Ministry of Defence, in 1992.[27] Professor Oxburgh argued that the objectives of Western military forces in the post-Cold War period will be more easily achievable if they have a technological edge. But he believes that such an edge is always relative and must be maintained by constant innovation. Indeed, he suggested that the Cold War might arguably be seen as a contest between East and West in the development of military technology and, further, that a technological edge might be a weapon in itself:

> . . . Certainly towards the end of the Cold War the United States selectively exposed the Soviet technological community to US achievements in space technology, presumably with the policy objective of encouraging in the Soviet Union a conviction that they were committed to a technological race that they could not win . . .

Professor Oxburgh stated that he was unsure of the success of the policy, but continued:

> . . . it seems likely that the conspicuous possession of an edge in military technology, and a vigorous determination to retain it, could even in the future be an important element in deterrence.

By 1993 the *SIPRI Yearbook* was announcing that this was exactly the policy of the United States. In a chapter entitled 'Military technology and international security: the case of the USA', the view was taken that:[28]

> During the year, a consensus emerged that the USA should make a major effort to retain the technological supremacy demonstrated in the war against Iraq . . .

Technology-denial regimes (including arms control) were said to be seen by many US policy-makers as 'an important corollary' of the aim of maintaining supremacy. Thus crucial technologies were to be stigmatised as indiscriminate, offensive, destabilising and so on in order to promote norms of denial. Interestingly also, according to the authors:

> Promoting such norms without adversely affecting US security policy is an explicit goal of an initiative to develop and deploy 'disabling' or 'non-lethal' weaponry . . .

Yet the globalisation of technology continues apace,[29] and even if the policies of the United States and other industrialised countries deter military adventure against themselves, Third World countries may be much more concerned about military threats from their own near neighbours. It is surely the case, for example, that India moved to obtain a nuclear capability largely because China had done so, and Pakistan certainly was strongly motivated to initiate its nuclear programme because of India's nuclear capability.[30] And it cannot be argued that major Third World countries are unaware of the move towards non-lethal weaponry in the industrialised nations.[31] Thus Western military technological innovation could end up serving more as a model for the world-wide introduction of new technology rather than as a means of preventing such a process. The unpleasant consequences of this kind of dynamic will be considered in the next chapter.

. 3 .

THE INHUMANE WEAPONS CONVENTION

The 1980 United Nations Convention on Prohibitions or Restrictions on the use of Certain Conventional Weapons which may be Deemed to be Excessively Injurious or to have Indiscriminate Effects, otherwise known as the Inhumane Weapons Convention (IWC), received its first review by the state parties in September 1995.[1] The Convention as a whole consists of an umbrella agreement on procedures and a series of specific protocols on weapon systems. As originally agreed in 1980, there were three protocols, none of which dealt with the issue of laser weapons. Yet it is striking that, in the run-up to the review, the International Committee of the Red Cross (ICRC) pinned considerable hopes on the possibility of incorporating hand-held laser weapons in the IWC, with the aim of reaching agreement on measures of control before the proliferation of these blinding weapons gets completely out of hand.

In a report on the Convention produced by the ICRC in the year prior to the review, 'blinding weapons' were the first of the specific weapon systems considered for possible additional protocols.[2] In the view of the ICRC:

> . . . Given the advanced stage of development of hand-held versions of this type of weapon, with the real possibility of their appearance on the battlefield in the near future and subsequent proliferation . . . it is essential that the Review Conference takes this last opportunity to take whatever preventive action is possible.

As usual, the ICRC's position was based on thorough analysis. Four meetings of experts had been held because of the reports that certain types of laser weapon could cause incurable blindness. The experts

. 29 .

pointed to some extremely worrying characteristics of hand-held laser weapons. The weapons then available for training purposes simply clipped on to rifles and could easily be designed to damage eyes. More powerful systems are now under development and could become increasingly cheap to produce. The systems under development for both anti-personnel and anti-sensor purposes would blind people permanently if they were hit from up to a distance of one kilometre, and even the battery-powered system would allow for a large number of shots. Protection by special goggles would be difficult because the lasers would operate across a wide range of wavelengths and therefore goggles designed to protect against some wavelengths would be useless against others.

These concerns have been discussed in the specialist literature. Rupert Pengelley, for example, wrote an editorial in *International Defense Review* entitled 'Wanted: a watch on non-lethal weapons', which began by pointing out the ICRC's particular concern over blinding weapons.[3] *Laser Focus World* carried a 'Washington Report' headed 'Public debate needed on laser weapons' which noted that:[4]

> A few years ago, *The New York Times* quoted a retired Army general as saying that blinding a soldier was better than killing one because more people are needed to take a blind soldier off the battlefield and care for that person . . .

The difficulty of balancing such claims of 'military necessity' against those of 'intolerable encroachment of humanity' bedevils the whole problem of developing international law in this area.[5]

The ICRC report was particularly concerned with the medical and psychological problems which would result for people who were blinded by such weapons. It suggested that if laser weapons were ever to become commonly used, serious eye damage could account for a quarter to a half of all battlefield casualties. Many other casualties would be the victims of repressive regimes, terrorists or criminals as proliferation inevitably proceeded. The ICRC report also pointed out that a number of approaches are available to the international community for dealing with this dangerous potential development – from banning the use of specific weapons or certain specified uses of a particular weapon, through to limits on the production and proliferation of types of weapon system. To understand how difficult this might be we need to look more closely at the Inhumane Weapons Convention itself and at how it came to take its present form.

The Inhumane Weapons Convention

The Inhumane Weapons Convention is part of that section of international law which deals with the regulation of armed conflict in order to mitigate human suffering. As Kalshoven pointedly noted in his review of modern constraints on the waging of war, written for the ICRC, some hard questions need to be asked about this whole body of law before engaging in the details of regulation.[5] Firstly, it has to be asked whether 'humanising' war is a credible aim, given that it appears to be contradictory to the whole essence of war. In Kalshoven's view the answer is that the humanitarian law of armed conflict does not intend to make war a humane activity; rather it attempts to restrain the belligerents and provide some essential protection for those involved. Secondly, Kalshoven suggests that it is necessary to ask if such an aim is, in fact, desirable. Could the effort to make war less inhumane deflect us from the real task of attempting to abolish warfare? Kalshoven argues that we must accept that warfare will not come to an end quickly. Surely, he suggests, it is presumptuous to argue that no just cause will arise in the future for which recourse to war will be necessary. Finally, he considers it necessary to ask whether the application of humanitarian law might not prolong suffering by extending the duration of a war which could have been ended more swiftly by use of greater force? This, he suggests, is an argument that may sound fine in theory but once thought about practically – by asking what an all-out war would be like – can be seen to be 'as untenable as it is abhorrent'. In short, while the idea of international humanitarian law designed to mitigate suffering in armed conflicts may seem strange at first sight, it is, in fact, sound.

Though the history of attempts to moderate warfare is a long one, modern efforts, developing as the industrialisation of warfare has increased over the last 150 years, are commonly held to rest on two bases: the Law of Geneva and the Law of The Hague. These two bodies of law have subsequently merged with current views on the protection of human rights, but it is still useful conceptually to think of:[6]

> . . . Geneva Law [which] constitutes a body of humanitarian law concerned with the treatment and protection of those who are *hors de combat*, civilians or otherwise exempt from treatment as combatants.

and:

> The law with regard to means and methods of conducting actual military operations in armed conflict [which] is generally known as the Law of The Hague.

Their titles derive from the development of the Red Cross movement in Geneva following Henri Dunant's experiences at the Battle of Solferino in 1859, and the Peace Conferences at The Hague at the turn of the last century.

The Law of Geneva has been developed much more substantially during this century than the Law of The Hague, with major developments of the law protecting the victims of warfare following both the First and Second World Wars. According to the Red Cross, the main elements of international humanitarian law are currently: the four Geneva Conventions of 1949; the two Additional Protocols of 1977; the Hague Convention of 1954; and the 1980 United Nations Inhumane Weapons Convention[7] (Table 3.1). The status of these agreements within the international community is also clearly reflected in the fact that in 1993 181 states were party to the Geneva Conventions of 1949, but to the Inhumane Weapons Convention were just 36.

Table 3.1 The main instruments of international humanitarian law*

Date	Title	Description
1949	The four Geneva Conventions	Protect the sick and wounded members of armed forces on land and at sea, and protect prisoners of war and civilians in wartime.
1977	The two Additional Protocols	Update the protection offered in the 1949 Conventions, adapt it to new forms of warfare (such as guerilla warfare) and enhance protection of the victims of international and non-international conflict.
1954	The Hague Convention	Protects sites and items of cultural importance from attack, misuse or abuse in wartime.
1980	The Inhumane Weapons Convention	Bans or restricts the use of certain conventional weapons which cause indiscriminate or excessive injury.

* From reference 7

The Law of The Hague is, in fact, pre-dated by the Lieber Code on the conduct of war, issued during the American Civil War, and the St. Petersburg Declaration of 1868, 'Renouncing the Use, in Time of War, of Explosive Projectiles Under 400 Grammes Weight'. While the application of international humanitarian law to specific weapon systems has developed haltingly since that time, the line of reasoning of the International Military Commission that met in St. Petersburg at the invitation of the Russian government still bears consideration today. They argued as follows[5] (Table 3.2): that the progress of civilisation required the alleviation of the calamities of war; that the only legitimate object in war was to weaken the forces of the enemy; to do this it was sufficient to disable the greatest number of men; that this objective would be exceeded by the use of weapons which uselessly aggravate suffering; thus the employment of such weapons would be contrary to the laws of humanity and the explosive projectiles in question should be banned.

Table 3.2   Arguments put forward by the St. Petersburg Military Commission*

1. The progress of civilisation should have the effect of alleviating as much as possible the calamities of war.

2. The only legitimate object which states should endeavour to accomplish during war is to weaken the military forces of the enemy.

3. For this purpose it would be 'sufficient to disable the greatest possible number of men'.

4. This object would be exceeded by the employment of arms which uselessly aggravate the suffering of disabled men or render their death inevitable.

5. The employment of such weapons 'would, therefore, be contrary to the laws of humanity'.

6. As in the eyes of the Commission the projectiles at issue met this qualification, it remained . . . to fix 'the technical limits at which the necessities of war ought to yield to the requirements of humanity'.

* From reference 5

It is noteworthy that the St. Petersburg Declaration also looked forward to future scientific developments in stating that:

> . . . Parties reserve to themselves to come hereafter to an understanding . . . of future improvements which science may effect in the armament of troops, in order to maintain the principles which they have established, and to conciliate the necessities of war with the laws of humanity.

This they did, of course, when 'dum-dum' bullets were also banned at the First Hague Peace Conference of 1899, and it is against that background of international law that we should consider the Inhumane Weapons Convention itself.

The umbrella Convention consists of a preamble and 11 articles

(Appendix 1). Its long title implicitly suggests that the text does not deal with such weapons of mass destruction as nuclear, chemical or biological arms but rather with what have been termed 'dubious' weapons.[5] As will be clear from the last section of this chapter, the Convention and protocols proved very difficult to negotiate and what resulted was modest in the extreme. The articles of the Convention deal with matters such as the scope of the agreement and the mechanism of review, in contrast to the protocols, which deal with specific weapon systems.

Article 1, with its reference back to the Geneva Conventions of 1949 and the first Additional Protocol of 1977, makes clear that the IWC is applicable to international armed conflicts, including wars of national liberation such as those which flared up during the period of decolonisation, and not to intra-state conflicts such as those which have so plagued the period since the end of the Cold War. The reason that the Convention was separated from the specific protocols was the uncertainty over whether individual states would be prepared to accept even what was agreed. Thus Article 4 has an elaborate mechanism whereby states must accept any two of the three agreed protocols (Article 4.3). The arrangements for entry into force, set out in Article 5, were quite quickly met and the Convention became operative in 1983. The rather complex arrangements for the application of the Convention to state parties confronted with situations where there are other states or non-state groups involved in conflicts are set out in Article 7. The provisions in regard to state parties facing other states are weighted towards application of the Convention, but the situation in regard to non-state groups is considerably more complex.

Article 8 deals with review and amendment of the Convention and protocols. Since negotiations were so difficult, it is not surprising that this article reflects a compromise between those who had wanted a more extensive original agreement and those who had sought to restrict it.[5] Article 8.1 allows for amendments to be proposed and for a conference to be held to consider any proposal, but only if a majority, then not less than 18, of the state parties so agree. On the other hand, Article 8.2 allows for the unrestricted proposal of 'additional protocols related to other categories of conventional weapons not covered by the existing annexed Protocols'. Article 8.3 provides the means for a state party to request the convening of a review conference after ten years if no amendment conference has been held. The review conference is empowered to cover the scope and operation of the Convention and to consider new protocols.

While the Secretary-General of the United Nations (the Depository of the Convention and protocols) has a number of designated tasks, commentators have noted the obvious omissions that weaken the Convention as a whole. There are few means for encouraging implementation or fact-finding, or for designating responsibility for what is done or not done in a conflict. This overall weakness also shows up clearly in each of the protocols. Protocol I, on Non-Detectable Fragments, consists of just one short sentence and completely prohibits use of any weapon which has the primary effect of causing injury with fragments which are not detectable by X-ray. It has been justly and heavily criticised for addressing a problem that did not actually exist. There were rumours of the development of plastic mines intended primarily to embed fragments undetectable by X-ray in victims' bodies. These turned out to be untrue but that has not prevented the development of plastic mines ostensibly for other reasons, such as non-detectability when emplaced in the ground. Given the limited value of Protocol I, it can be understood why Article 4.3 of the Convention required acceptance of two or more of the protocols by state parties. Acceptance of Protocol I alone would, as one commentator[5] remarked, 'have been all too absurd'.

Protocol II is concerned with 'Prohibitions or Restrictions on the Use of Mines, Booby Traps and Other Devices'. We shall deal with the consequences of its inadequacies in the next chapter, so a detailed review of the protocol itself is necessary here. Article 1 of the protocol deals with the scope of application. Use on land of mines, booby traps and other devices (defined in Article 2) including 'mines used to interdict beaches, waterway crossings or river crossings' is covered, but use of anti-ship mines at sea or on inland waterways is not. As pointed out by Ove Bring, a legal adviser to the Ministry of Foreign Affairs in Sweden, modernisation of the old laws regarding sea mines is probably also necessary.[8]

According to the definitions given in Article 2 a mine is:

> . . . any munition placed under, on or near the ground or other surface area and designed to be detonated or exploded by the presence, proximity or contact of a person or vehicle . . .

and a remotely-delivered mine is any such device delivered, for example, by artillery or aircraft. A booby-trap is defined as:

> . . . any device or material which is designed, constructed or adapted to kill or injure and which functions unexpectedly when a person disturbs or approaches an apparently harmless object or performs an apparently safe act.

Finally, other devices are defined as: 'manually-emplaced munitions and devices designed to kill, injure or damage and which are activated by remote control or automatically after a lapse of time'. This all appears commendably clear and straightforward.

Article 3 then sets out the general restrictions on the use of such munitions. Commenting on the protocol soon after it was agreed, Jozef Goldblat noted that:[9]

> The use of mines, booby traps and other devices against the civilian population as such, or against individual civilians, is prohibited in all circumstances, whether in offence or defence or by way of reprisals . . .

He pointed out that:

> . . . Also prohibited is the indiscriminate use of all these devices against military objectives in conditions which may be expected to cause incidental loss of civilian life, injury to civilians or damage to civilian objects, excessive in relation to concrete and direct military advantage anticipated . . .

Article 3.4 enjoins parties to take 'All feasible precautions' to protect civilians. Article 4 restricts the use of these munitions in towns, villages and other concentrations of civilians. However, the prohibition only applies if 'combat between ground forces is not taking place or does not appear imminent', and even then may be overridden if the munitions are either placed in close proximity to a military objective or measures to protect civilians, such as warning signs, are used.

The protocol defines a military objective (Article 2.4) as :

> . . . any object which by its nature, location, purpose or use makes an effective contribution to military action and whose total or partial destruction, capture or neutralisation, in the circumstances ruling at the time, offers a definite military advantage.

It is prohibited, in Article 5, to use remotely-delivered mines unless they are placed in an area which is a military objective or one which contains military objectives, unless the location of the mines can be accurately recorded or an effective neutralising mechanism is incorporated into each mine. It is also necessary to give effective advance warning of the use of such mines if they may affect the civilian population 'unless circumstances do not permit'. Article 6 prohibits, in all circumstances, the use of booby-traps and defines a wide range of these. A representative example would be a device associated with:

> . . . children's toys or other portable objects or products specially designed
> for the feeding, health, hygiene, clothing or education of children.

Article 7 is concerned with recording the location of minefields, with the retention of such information and with provisions for its use immediately after the cessation of hostilities. The problem here is that the Article refers specifically to records related (Article 7.1 (a)) to 'all pre-planned minefields' and thus opens up the possibility that failure to record hastily-emplaced minefields would be acceptable. Since Article 5 refers directly to the location of remotely delivered mines being 'accurately recorded in accordance with Article 7.1.(a)', the same argument/exemption would also apply to these mines.

Article 8 gives some limited protection for United Nations' forces and missions and Article 9 sketches some measures of international co-operation in the removal of these munitions. A short technical annexe is concerned with the recording of the location of minefields and other munitions.

Protocol III has two Articles dealing with 'Prohibitions or Restrictions on the Use of Incendiary Weapons'. Article 1.1 provides a definition and examples of what these are considered to be. An incendiary weapon means:

> . . . any weapon or munition which is primarily designed to set fire to
> objects or to cause burn injury to persons through the action of flame,
> heat, or a combination thereof, produced by a chemical reaction of a
> substance delivered on the target.

Examples given in the text of the protocol include flame-throwers, shells, rockets, mines and bombs. Article 2 actually specifies in its title that it applies only to civilians and civilian objects, and some of the exceptions in applications, for example in the final paragraph (2.4), are such as to render the restrictions of little practical value. It has, however, been argued that at least Article 2.2 makes it clear that fire-bombing of cities from the air, in order to destroy military objectives, is illegal.

The difficulties of reaching agreement were so marked that only a 'Resolution' could be put together on the question of small-calibre weapons. The threat was of a new range of bullets which might have similar effects to those banned in 1899. An appeal was made to governments 'to exercise the utmost care . . . so as to avoid an unnecessary escalation of the injurious effects of such systems', pending further research and international discussions. Suggestions put forward for improving the IWC at its 1995 review showed clearly its present weaknesses.

Improving the IWC

While the general rules of international humanitarian law could be argued to apply to the kinds of internal warfare that have arisen since the end of the Cold War, many people have suggested that the IWC should be made to apply specifically to such conflicts by amendment of Article I of the Convention. It has also been argued that there are a variety of ways in which implementation of the Convention could be improved, for example: by incorporation of legal advisers into military operations in a more effective manner; by much more rigorous training of armed forces in regard to the requirements of the Convention; and by incorporation of the Convention into national legislation. Wider-ranging suggestions included one that mechanisms of international fact-finding be incorporated into the Convention or even that a special supervisory body be established. There were also suggestions that there should be incorporation of liability and criminal sanctions, adjudication mechanisms and even a role for the Security Council in the imposition of penalties following violations of the Convention.

In regard to the specific protocols, most attention was directed at Protocol II because of the dreadful impact of the widespread use of anti-personnel mines (see Chapter 4). The ICRC was so concerned at this that it organised an international symposium on the subject at Montreux in April 1993, and then a follow-up meeting of military experts in January 1994. We shall consider the general outcome of these meetings in the next chapter and concentrate here on the ICRC's proposals for what might have been done regarding Protocol II at the review conference.[2] The proposals fell into two broad groups: prohibitions on certain types of mine and prohibitions on how mines are to be used. Clearly, however, any additional provisions in this humanitarian law would have to be supplemented by further measures of arms control if they were to be effective, because anti-personnel mines are presently so cheaply and widely available.

The most straightforward proposal was to prohibit the use of anti-personnel mines altogether, an act which would clearly demonstrate the unacceptability of such weapons and simplify verification. Great care would have to be taken in defining the banned systems so that regulation could not be easily evaded, and a strict ban on production and stockpiling, with effective verification, would be needed. While it might be argued that a ban on the use of anti-personnel mines lacking self-destruct mechanisms might be more easily agreed, verification would be more difficult and the end result could still be the emplacement of large numbers of mines with long-term lethality if the self-destruct mechanism were not 100 per cent effective.

Theoretically, there were several ways in which the restrictions on the use of anti-personnel mines could be improved. Proposals included tighter requirements on the recording of minefield positions, declaration of this information after the close of hostilities, and clear provision for allocating liability in clearing minefields. However, even if amendments could have been agreed, there would still be considerable reservations about the widespread observance of detailed new regulations in the kind of disordered internal conflict we are seeing at present.

The ICRC also suggested that the review conference discuss the question of small-calibre weapons.[2] It argued that the considerable research undertaken since 1980 confirmed that early turning or break-up of the bullet in the body results in greater energy transfer and therefore greater damage. These problems result from the poor stability of the bullet and the materials and construction used. Some states had already taken action to correct the design of their bullets and the Swiss offered to put their testing facilities at the disposal of other states, to assist better understanding of resistance to fragmentation. The ICRC suggested that the review conference should consider how best to take advantage of these developments in order to improve the law.

One weapon system that has been increasingly used in recent years is the cluster bomb, in which many small bomblets are released in a package. The bomblets are small but can cause great damage when they explode. Unfortunately, there is apparently a very high failure rate of such bomblets on initial release. Thus a typical package of 360–650 bomblets with a common failure rate of 40 per cent would leave 144–260 highly unstable bomblets on the ground where they could cause significant injury to anyone unlucky enough to set one off. To make matters worse, use of this type of weapon has moved well beyond its original purpose of attacking airstrips. They are now frequently considered as area-denial weapons, rather similar in some ways to mines. However, the clearance problem is even worse because of the instability of the unexploded portion of the bomblets released. The only real solution is to attempt to destroy the unexploded weapons, yet even this is extremely dangerous. Manufacturers have responded to the unexpectedly high failure rates by installing self-destruct mechanisms in newer models (clearly the unexploded bomblets can be a threat to friendly as well as hostile forces if the area of operations changes). The ICRC suggested that the review conference should 'seriously consider making such a measure mandatory'. That would certainly prevent a repetition of the problem so long as the self-destruct mechanism itself was highly reliable.

The views of the ICRC on hand-held laser weapons were referred to at the start of this chapter. For such weapons, a variety of proposals were put forward. It seemed probable that only a tough and tightly-worded approach would effectively stem their proliferation. One proposal, for example, would have banned blinding weapons, using wording such as,'weapons may not be used against persons with the primary intention or expected result of permanently damaging their eyesight',[2] and would have backed this up with a duty to take precautions against accidental damage to eyesight, and prohibitions and restrictions on production and proliferation in associated arms control agreements.

As indicated previously, the 1995 Review Conference did not make many advances in extending and strengthening the IWC. Yet the question surely is, 'Why were we still discussing the failure of the IWC in 1995?', a decade and a half after the initial failure to produce an effective international agreement. To answer that question we need to know more about the history of the negotiation of the IWC.

The Negotiation of the IWC

It is very important to understand that while the IWC is a disappointingly ineffective legal instrument, it did not result from some small-scale *ad hoc* initiative. Rather it resulted from a long-drawn-out process of large-scale international effort. This was charted carefully by the Stockholm International Peace Research Institute (SIPRI) in at least four *Yearbook* chapters[10,11,12,13] and two books.[14,15,16] The range of weapon systems considered was extensive (Table 3.3) and there were numerous international meetings leading up to the attempt to conclude the Convention (Table 3.4).

Following the massive scale of the war crimes committed in the Second World War, there was naturally an initial concentration on upgrading the Geneva Conventions but, following the Korean War, the ICRC did give some consideration to restrictions on specific weapon systems.[15] Then the terrible carnage of the Vietnam War led to part of the reaction against the war being directed against many of the modern weapons used by the Americans. The international tribunal named after Bertrand Russell heard testimony from experts on such weapons and helped to focus public attention. The issue was then taken up in the United Nations and again by the ICRC. The Diplomatic Conference which was working to upgrade the 1949 Geneva Conventions

Table 3.3   The range of weapons systems considered for restriction in the 1970s*

Nuclear weapons

Biological weapons

Chemical weapons

Geophysical weapons

Incendiary weapons

Small-calibre weapons

Fragmentation weapons (eg cluster bombs)

Delayed-action weapons (eg mines)

Electric, acoustic and electromagnetic wave weapons (eg light, lasers, microwaves)

---

* From references 14 and 15

Table 3.4   Some international actions taken in the period prior to the agreement of the Inhumane Weapons Convention (IWC)*

October 1972	UN Secretary-General reports on *Napalm and Other Incendiary Weapons and all Aspects of their Possible Use.*
November 1973	UN Secretariat publishes survey of *Existing Rules of International Law Concerning the Prohibition or Restriction of Use of Specific Weapons.*
November 1973	XXIInd International Conference of the Red Cross welcomes experts' report on *Weapons that may Cause Unnecessary Suffering or Have Indiscriminate Effects.*
February–March 1974	Diplomatic Conference on the Reaffirmation and Development of International Humanitarian Law Applicable in Armed Conflicts establishes *ad hoc* Committee on conventional weapon restrictions.
September–October 1974	ICRC Conference of Government Experts on Weapons that may Cause Unnecessary Suffering or Have Indiscriminate Effects held at Lucerne.
January–February 1976	Second Conference held at Lugano.
March–June 1977	Diplomatic Conference on International Humanitarian Law in Geneva agrees two Additional Protocols to 1949 Geneva Conventions and calls on UN to convene a conference on conventional weapons.
1979–1980	Two sessions of the UN Conference on Prohibitions or Restrictions of Use of Certain Conventional Weapons which may be Deemed to be Excessively Injurious or to have Indiscriminate Effects held in Geneva and agree IWC.

---

* From references 15 and 16

established an *ad hoc* committee to look at possible restrictions on specific weapons and, in its final session in 1977, the Conference called on the UN to hold a special conference on the subject.

Meanwhile, the UN Secretary-General published a report on napalm and other incendiaries in 1973 and, in the same year, the ICRC published an experts' group report on weapons that could cause unnecessary suffering or have indiscriminate effects. According to SIPRI, the major military powers had not been well represented in the experts' groups which produced these reports. They did, however, attend in force at two subsequent conferences held in Lucerne in 1974 and Lugano in 1976. According to SIPRI:[15]

> ... the resulting reports (ICRC 1974, 1976) tend to give more emphasis to the military utility, or even 'necessity', of some of the weapons which in earlier reports had been regarded as possible candidates for restrictive measures.

SIPRI's view, on the other hand, was that weapons such as napalm and phosphorus could be so cruel as to their warrant banning even if they had military use. This tension between military and humanitarian considerations is present throughout the SIPRI reports[10,11,12,13] through the 1970s and, clearly, the military considerations were given greater weight in the ineffective legal instrument which resulted in 1980.

Of course, that is not to argue that the negotiators deliberately set out to achieve such a result. As Röling and Suković pointed out, at the first session of the Diplomatic Conference on the Reaffirmation and Development of International Humanitarian Law Applicable in Armed Conflicts, in 1974, some delegates argued against confining work just to conventional weapons and wanted consideration also to be given to nuclear and other weapons of mass destruction. However, by the third session of the conference, delegates realistically restricted the topics to:[14]

> ... four categories of conventional weapons: incendiary weapons, small-calibre high-velocity bullets, fragmentation weapons and delayed-action weapons.

These clearly represent the topics covered, in part, by the three Protocols and the Resolution eventually agreed in 1980. The interesting point is how the legal regulations for each of these topics came to be narrowed down, and why.

Agents for incendiary weapons can take a variety of forms. At the time of the discussions of the IWC, napalm was perhaps the most widely known. Since petrol burns too rapidly for effective use as an incendiary weapon, napalm was made by using additives which cause it to burn in sticky globules for a longer time and at a higher temperature. Strategic use of incendiaries to create firestorms, as in the Second World War, is indiscriminate but tactical use of napalm can also be indiscriminate in its effects. Burn injuries caused by such weapons can obviously be extremely painful and difficult to treat. Between the World Wars, considerable attention was paid to incendiary weapons at the Geneva Disarmament Conference in 1932–33. The draft Convention[14] 'would have explicitly forbidden the use of projectiles specifically intended to cause fire, and appliances designed to attack persons by fire'. A similar approach, of straightforward prohibition, informed the early post-Second World War work of the ICRC in the 1950s.

The general approach adopted in the initial stages of discussion of the IWC followed this line closely. Indeed, it was 'largely modelled upon Article 48 of Part 4 of the British draft disarmament Convention of 16 March 1933'.[14] A revised version of the proposal, sponsored by a wide range of states from Algeria to New Zealand and Sweden to Zaire, is set out in Table 3.5. Since such weapons violated many of the principles of law, the idea was to have a broad ban with some specific exceptions if these were unavoidable.

Table 3.5 A proposed protocol on incendiary weapons*

---

1. Incendiary weapons shall be prohibited for use.

2. This prohibition shall apply to:

   the use of any munition which is primarily designed to set fire to objects or to cause burn injuries to persons through the action of flame and/or heat produced by a chemical reaction of a substance delivered on the target. Such munitions include flame-throwers, incendiary shells, rockets, grenades, mines and bombs.

3. This prohibition shall not apply to:

   (a) munitions which may have secondary or incidental incendiary effects, such as illuminants, tracers, smoke or signalling systems.

   (b) munitions which combine incendiary effects with penetration or fragmentation effects and which are specifically designed for use against aircraft, armoured vehicles and similar targets.

---

* From reference 14

Comparison of Protocol III of the IWC (Appendix I) with the proposal set out in Table 3.5 shows just how far from a prohibition resulted from the 'hard battle' at the conference.[5] Whereas the proposal starts with a general ban and applies this to a list of weapon systems, in the protocol, as Kalshoven noted:[5]

> . . . This list of examples is not, however, of material significance . . . as none of the incendiary weapons included in the list has been made the subject of separate regulation, whether prohibition or restriction. The same applies with even greater force to napalm, which is not even mentioned in the list.

While there may be scope for disagreement as to the protection actually afforded by Protocol III, there can be little doubt that it is a far cry from what was intended by those who argued for a broad prohibition of such weapons on the grounds of their indiscriminate and unnecessarily excessive effects. The reports on the subject of conventional weapons regulation in the *SIPRI Yearbooks* during the 1970s provide an insight into why this happened. In its report on the Lucerne meeting in 1974 SIPRI argued that it was clear that incendiary weapons were the most likely systems to be banned in the negotiation process under way.[11] This was understandable given the amount of attention that had been paid to this subject, and particularly to the use of napalm. Yet the SIPRI authors noted that:[11]

> . . . considerable effort was expended on questions of definition and classification. Though it was not stated explicitly, the purpose of these exercises was not to deepen the level of scientific study but to limit the scope of possible prohibitions.

The reasons why major military powers might adopt such a restrictive attitude had already been set out by SIPRI. In their 1973 report, after going into considerable detail on the growing use and sophistication of incendiary weapons, the authors argued that there were two overwhelming military demands at that time in regard to weapons for battlefield use:[10]

> The first is for weapons which *permanently incapacitate even when only non-vital areas of the body are hit*. . . . The second military demand is for *weapons which cover an area which may include the target, rather than hitting the target directly* . . .

These demands existed despite the fact that the first requirement appears to be directly contrary to the requirement that unnecessary suffering be limited, and the second to the requirement that indiscriminate effects be restricted. For example, 'Napalm bombs combine area characteristics with high incapacitating power' and would therefore have military attractions.[10] Thus there was opposition to a prohibition from the major military powers despite the fact that such weapons were cheap and simple to manufacture, and the necessary raw materials were widely available around the world.[14] Unfortunately, in these circumstances proliferation of such weaponry became a virtual certainty.

In its 1979 report, SIPRI set out what it thought should be achieved by the UN Conference in regard to specific weapon systems (Table 3.6). By the time of their 1980 report, the first negotiation session (September 1979) had demonstrated that such results were unlikely. SIPRI criticised the limited scope of the umbrella Convention proposed by the UK and Netherlands which:[13]

> ... was remarkable since it ignored both the fact that many of the uses and abuses of the various weapons under discussion have occurred in internal conflicts, as well as the progress of international law in this respect since 1949 – in particular the 1977 Additional Protocol.

Yet internal conflicts were only included in the scope of the IWC in 1995.

Table 3.6   Possible aims of the IWC negotiations*

(a)   a ban on the use of nuclear weapons;

(b)   a ban on incendiary weapons, including white phosphorus and new incendiary agents;

(c)   a ban on bullets which tumble or break up within 150 mm of tissue . . . and on multiple-projectile bullets;

(d)   a ban on fuel-air explosives;

(e)   restrictions on the use of fragmentation weapons within specified zones of inhabited areas;

(f)   a prohibition on mines which cannot be located and disposed of by known means;

(g)   an obligation on those who employ munitions such as mines to ensure their removal at the end of hostilities; and

(h)   restrictions on the use of mines, booby-traps and other delayed-action munitions in order to minimise civilian casualties.

---

* From reference 12

The SIPRI report, in passages which bear re-reading in detail since we will need to engage in a continuing review of the IWC, contrasts the views of those who desire an effective mechanism for dealing with new military technology and those who wish to avoid international restriction on national freedom of action in military affairs. Thus the authors argued:[13]

> . . . 'review' can mean anything from a reluctant undertaking to meet, say, ten years after the entry into force of some minimal agreement . . . to a much more explicit requirement to meet, at shorter intervals, to continue the substantive business of outlawing inhumane and indiscriminate weapons . . .

In fact, a year before the conclusion of the negotiations for the IWC, the SIPRI authors had concluded that in taking the pragmatic approach of focusing on specific weapons, 'the conference finds its time taken up with trying to accommodate the military interests of each state and military service'. They suggested that:

> . . . What is likely to emerge is not a clear normative statement about superfluous injury and indiscriminate effects – followed by a list of weapons to which these criteria apply – but a modest set of rules permitting the use of any weapon which any state can claim is needed for a particular task.

Reflecting on his experience of the process whereby the original hopes for effective restrictions were watered down to the final IWC text, Eric Prokosch wrote in 1995 that the meagre results:[16]

> . . . can be attributed to the willingness of most participants to defer to their military experts' claims that the weapons under consideration were very useful, not very harmful, or not inherently indiscriminate.

The points are reinforced with alarming examples from his own encounters at the various conferences in the series.

Finally for our consideration of 'non-lethal' weapons, it is important to note that discussions such as those at Lucerne in 1974 had even covered such weapons of the future as lasers, microwaves, infrasound and light-flash devices.[11] However, before returning to the general problem of non-lethal weapons we will consider, in the next chapter, the massive human disaster which has resulted from the failure to deal adequately with delayed-action weapons – that is, mines – in the IWC. This should leave no doubt about potential future catastrophes if we avoid considering what a mix of successful, cheap, new, military technology and political opportunity for its use can bring.

.4.

FROM DEFENSIVE MINES TO ANTI-PERSONNEL DESERTS

The history of warfare may be seen as a series of developments of offensive measures and defensive countermeasures. Thus the introduction of tanks, which had the potential to break the stalemate of the kind of trench warfare which characterised the First World War, led to the development of anti-tank landmines prior to the Second World War. Landmines were used tactically in order to counter the tank, and although time-consuming and labour-intensive to emplace, were deployed in huge numbers during the Second World War.[1]

This massive use of mines in 1939–45 led to enormous problems of clearance after the war. Poland was the most heavily mined country with about 80 per cent of the land affected, and 600,000 hectares very severely mined. Between 1945 and 1956 some 10,000 men cleared about 15 million mines. These engineers had the advantage of 17,000 maps of the minefields supplied by the Soviet Union and did use metal detectors, but the preferred method was slow and tedious probing with pointed rods. Despite the considerable effort, and success, 300 specialists were still being employed during the 1980s to clear thousands of mines, and other explosive remnants of the war. Hundreds of people have been killed attempting to clear such debris since 1945 and some civilians, particularly children, continue to be killed each year. Similarly, in France, 13,000 remnants were removed in 1978 alone and in the early 1980s, 80 disposal experts were still continuously employed in clearance. However, little attempt has been made to clear the many millions of mines laid in North Africa during the Second World War. In Libya, between 1945 and 1975, some 4,000 people were killed and 30 to 40 are said to be still killed each year by ordnance first emplaced five decades ago.

Although the initial impetus for the development of landmines was to counter the tank, military developments and counteraction did not stop there. Anti-tank mines were large and contained some 10 kilograms of explosive[2] – they could be found, cleared and even re-used by an antagonist. Therefore, means were sought to protect mines, once laid, from disturbance. One method was to 'seed' minefields with smaller anti-personnel mines containing just a few grams of explosive. These mines could be set to explode if triggered by a person's foot or the displacement of a trip-wire.

After the Second World War vast military expenditures continued during the Cold War and, inevitably, military research and development produced new kinds of anti-personnel mine. During the Vietnam War particularly, the United States used very large numbers of cheaply-produced, remotely-delivered mines to deny land to enemy forces and to attack their supply lines. The scene was thus set for a modern, large-scale tragedy to be played out in the succeeding conflicts in various Third World countries.

Types of Landmine

Though anti-personnel landmines range in sophistication from the crude devices made by Third World insurgents through to high-technology weapons incorporating modern electronics, they all fall into four broad classes: blast; fragmentation; directional; and bounding.[3,4]

Among the roughly 340 anti-personnel landmine models produced in 48 different countries **blast mines** are the most common. These weapons are usually triggered when a victim steps on the mine. The explosive blast drives pieces of the mine, dirt, bits of footware etc. up into the person's body. If death does not result, amputation of a limb is frequently required. Examples of this type of mine are the Soviet-made PMN and PMN-2, gruesomely named 'black widow' mines because of their black thermo-plastic outer coating and devastating effects.

Fragmentation mines or 'stake' mines are laid above ground on stakes or attached to other supports such as trees and are usually activated via a trip-wire. The aim is to have the fragments of the mine ejected over as large an area as possible on activation. To this end, the casing often has a surface pattern indented or partly cut through or is surrounded by coils of partially cut-through wire to ensure that a larger number of small, pre-formed fragments disperse on explosion rather than fewer, larger, randomly-broken pieces. These smaller fragments obviously fly further.

Directional mines also rely on fragments of various kinds for their lethal effects. These fragments are located in front of an explosive charge and are thrown forward over a predetermined area when the mine is set off. The widely-copied US M18A1 *Claymore* mine contains some 700 steel balls which can be propelled over a range of 50 metres in a 60° arc about two metres high.

Bounding mines may rely partly on blast, but again are mainly dependent on fragmentation effects. The mine is located in a short barrel and is shot upwards when the device is first activated. When the mine reaches a certain height it explodes (for example, via the mechanism of an anchor cable pulling a fuse pin from the mine), scattering the fragments over a much wider area than if the mine had just exploded at ground level.

Obviously, mines can be triggered simply by pressure – when someone steps on the mine – or via disturbance of a trip-wire. More complex methods of activation are by remote control or by an electronic system. It has been argued that a mine such as the *Claymore*, under remote control by a soldier, is less dangerous than other types of mine and its use, for instance to defend static installations, should therefore not be such an immediate concern for arms control legislation.[5] Attachment of an electronic triggering device opens up many possibilities such as, for example, arranging for the mine to detonate only on the passage of a *second* potential victim. This would clearly be most disconcerting for an enemy force which had assumed an area to be safe when the first person passed unharmed. On the other hand, inclusion of an electronic system does allow relatively easy incorporation of a self-destruct or self-neutralisation mechanism into a mine.[6] While both mechanisms present problems – self-destruct because of the unexpected explosion and self-neutralisation because no-one can be sure it has actually happened – both have figured prominently in recent discussions of possible means of making landmines less dangerous to civilians.

Military operations today are ever more dependent on the speed of manoeuvre of armoured forces, and thus the original tactical defensive role of landmines remains important in standard military operations involving regular forces. However, the capability to deliver modern anti-personnel mines remotely, if indiscriminately, from aircraft in Vietnam did allow US forces to use these weapons in a much more offensive and strategic role, for example to interdict enemy supply routes. As the availability of cheap landmines has increased, so such offensive and strategic uses have been adopted by more irregular forces within internal wars. The US State Department described the process carefully in 1993:[7]

. . . the last two decades have seen an increasing use of landmines in non-traditional ways. Increasingly, guerilla and terrorist groups are using landmines to achieve political and economic, rather than military objectives.

An action and reaction process all too often then sets in, with insurgent groups attempting to achieve widespread disruption through the use of landmines and government forces in turn attempting to protect key elements of the infrastructure, such as roads and bridges, by using landmines. The result is to put in place what has been called 'a weapon of mass destruction in slow motion'. According to the view expressed in a study by the Arms Project of the US-based non-governmental organisation (NGO), Human Rights Watch, working in association with the Physicians for Human Rights:[4]

> . . . With this fundamental shift in role from tactical, battlefield weapon to theater-wide strategic weapon, landmines take on certain characteristics of weapons of mass destruction: they blight the land practically forever . . .

We shall now consider some of the disastrous consequences of deploying landmines.

Anti-Personnel Deserts

The countries which now suffer most from the consequences of landmine use are, of course, those which have experienced major warfare in the recent past. Obvious and well-known examples are Afghanistan, Angola, Cambodia and Mozambique, but the problem is worldwide. The US State Department estimates, from its 1993 survey, that landmines pose a threat to civilians in 56 countries and that about two-thirds of these countries need help with de-mining.[7] The survey suggests that there are between 65 and 110 million uncleared landmines worldwide and uses a figure of 85 million as a general working estimate. This figure includes mines of all types but anti-personnel mines predominate because of their very low cost. The survey shows emphatically that:

> . . . landmines are largely a rural Third World problem. Most of the uncleared landmines in the world today were laid in rural areas. Urban mining is the exception . . . .Mines are most commonly found in rural border areas, around rural infrastructure such as electric lines, water plants, and bridges, around military installations . . . and along roads. Terror mining of small towns and villages is common . . .

In short, it is the poorest of the poor who suffer most from this persistent plague.

Africa is probably the most mined region of the world, with a third of the countries in the continent having some problem with uncleared landmines. Between 18 and 30 million mines are emplaced in 18 countries. In the Middle East there are some 17 to 24 million mines, mostly in Kuwait, Iran and Iraq but also around Israel's borders. In total, eight countries in the Middle East region have problems with uncleared mines. Most of the 15 to 23 million mines in East Asia are located in South-East Asia. Eight countries there have problems with uncleared mines and the situation in Cambodia is one of the worst in the world. It is possible to make only a rough estimate, but the State Department suggests that four million mines may be emplaced there. In South Asia there may be anything between 13 and 25 million mines, mostly in Afghanistan, which the UN suggests has some 9–10 million uncleared mines. The State Department report states bluntly that 'Afghanistan is crippled by uncleared landmines'. In Europe there are thought to be from three to seven million landmines in 13 countries. These include some remaining from the Second World War but Europe is also experiencing the fastest increase in mining. This is happening in some of the republics of the former Soviet Union and in former Yugoslavia. Finally, in Latin America there are between 300,000 and a million uncleared mines in eight countries.

The Human Rights Watch and Physicians for Human Rights study makes a telling final point in its global survey of the problem. It presents detailed case studies of eight countries: Angola; Cambodia; El Salvador; Iraq; Kurdistan; Mozambique; Nicaragua; and Northern Somalia, and comments:[4]

> . . . most parties have used landmines deliberately against civilians, all have engaged in indiscriminate use in which the potential for severe harm to civilians was ignored, and few have ever taken precautions that could protect civilians . . .

As the report acknowledges, these countries are not party to Protocol II of the IWC on mines and in any case, these internal wars would not have been covered by the then current protocol. Yet, as the report also points out, 'all parties are bound by customary international law which unequivocally prohibits these practices'.

Again, the message would appear to be that arms control and disarmament measures are best applied prior to the deployment of any new weapons system. Once there is widespread deployment, it is very difficult, particularly in the chaotic conditions that pertain in a civil war, to bring restrictions to bear, especially if the specific restrictions are weak and unverifiable. Bearing that in mind, we will look briefly at two of the worst tragedies of modern landmining.

Afghanistan

According to the US State Department report, there are mines everywhere:[7]

> . . . on arable land and lowland grazing terrain, on footpaths, on all classes of roads, on the hillsides and mountainous grazing land, on hilltops and mountaintops, and in irrigation channels and canals in both urban and rural settings . . .

The primary responsibility for laying these mines rests with the forces of the former Soviet Union. The forces from this industrialised, northern country evolved a strategy of offensive mining in their war with the Mujahideen. Many of the mines were remotely delivered and are therefore not marked adequately. There are said to have been at least 20 types of Soviet anti-personnel mine found in the country, in addition to anti-personnel mines from five other countries and many anti-tank mine types.[4]

In December 1993 *Time* magazine reported that 35 teams of about 30 men each had succeeded in clearing 60,000 mines from around major cities such as Kabul, but noted that millions of acres of land remained to be cleared. Indeed, it quoted a Red Cross estimate that it could take 4,300 years to clear only 20 per cent of the country.[8] The human cost of these landmines is quite terrible to contemplate. It has been estimated that 200,000 people have died from mine explosions and another 400,000 have been injured.[4] A very large percentage of those injured by landmines in Afghanistan have required an amputation and, of course, amputees require further operations later in life.[1] Against a background of such consequences, after a war which lasted more than ten years, the problems of rehabilitating people who fled as refugees to Pakistan are immense.

Only detailed accounts can really convey the depth of the tragedy in terms that we can easily grasp. The Mines Advisory Group, a British NGO, gave two brief illustrations in a 1991 report.[9] Two farmers explained that they could not return to their village because not a single house remained standing and:

> . . . They could not plant crops because the land was covered in mines; attempts to continue farming had caused the loss of 12 men in the previous two years . . .

Others had been killed trying to clear the mines from a field near the village and after that they had all given up. The local commander explained that mine clearance was very difficult because the variety of different mines laid in the area necessitated expert assistance. He was not even sure that all the anti-tank mines laid on local roads by the Mujahideen had been successfully cleared.

At another village, situated in what must once have been a beautiful valley high in the wooded mountains, the scene was totally transformed:

> Now the forests are sparse, the land arid and the valley a scene of mindless destruction. The two irrigation canals . . . which had once fed over . . . 700 acres of agricultural land, are overgrown and in disrepair, the fields they watered unplanted . . .

Only four families remained out of more than 400 people who had once lived in the now-devastated village. The local commander tried to quantify the problem by suggesting that there could be 60,000 mines in the surrounding fields and hills. The biggest problem was how to collect wood for survival in an environment where two metres of snow and temperatures of $-20°$ C were not unusual in winter. Two villagers had recently died and one had been injured while attempting to gather wood. Another worry was that some of the remaining houses were known to be booby-trapped and others were suspected of being similarly rigged to kill anyone entering.

The report noted that these are just two of nearly 50 villages in the district and 'their plight is not an unusual one'. Moreover, the story is 'repeated throughout the area and in many such districts throughout Afghanistan'.

Cambodia

The Cambodian tragedy has continued for more than two decades.[4] It might even be dated from the early 1960s when US special forces teams began to make incursions from South Vietnam in order to lay mines.[10] It could be dated from the time of the massive, secret, US bombing campaign of the early 1970s but definitely from 1975, when the Khmer Rouge began their bloody reign of terror following the end of US involvement in the Vietnam War. Part of the Khmer Rouge mania was directed against Vietnam which eventually, in 1979, invaded Cambodia. This initiated what then became a war of the Vietnamese, and its puppet Cambodian government, against the Khmer Rouge and two other Cambodian factions. The war lasted until the end of the Cold War, and the commencement of the UN operations of the early 1990s, which were intended to re-establish peace.

According to the US State Department survey,[7] 'In no country in the world have uncleared landmines had such an enormous adverse impact as in Cambodia'. The length of the war was doubtless a factor here, allowing many mines to be laid over a lengthy period, but the severity of the situation also results from the way in which all parties to the war after 1979 decided to use mines. The report by Human Rights Watch noted that:[4]

> For all parties . . . the main purpose of laying landmines was to limit military operations by enemy forces: to deny opponents access to bridges, roads or strategic installations or to protect one's own forces from attack . . .

Mines were also used tactically, for example in operations in which:

> . . . Government troops . . . placed mines around the perimeters of enemy villages and then bombarded them with artillery fire so that the 'enemy' was forced to flee into the minefields . . .

The overall outcome was that there was little direct contact between opposing forces, and thus the Cambodian conflict may well have been:

> . . . the first war in history in which landmines have claimed more victims – combatants and non-combatants alike – than any other weapon . . .

Given the problems of clearing the massive numbers of mines which remain dispersed over half the country, the process of killing and maiming is set to continue well into the next century.

It is impossible to give a full account of the vast numbers of different types of mine present in Cambodia today. Certainly all four of the main classes of anti-personnel mine were used in the warfare there.[10] The commonest mine in the blast mine category is probably the PMN-2, originally manufactured in the Soviet Union. The Soviet-made POM-2 and POMZ-2M fragmentation mines are the most commonly found of this type. Chinese-made Type 69 mines are the most commonly found bounding mine. An example of the directional mines used in Cambodia is the MON-50 mine, originally manufactured in the Soviet Union. This has a 50-metre killing zone.

All parties to the 1979–91 Cambodian war were backed by outside powers and these influenced to some extent the different strategies and doctrines used by the military forces. However, a clear development was the deliberate increase in the destructiveness of the mine systems. One improvisation was to stack mines one on top of another. In an extreme form:[4]

> . . . an anti-tank mine called the TM-46 was placed above a standard
> pressure mine called the PMN with a Type 69 bounding mine at the
> bottom of the stack . . .

This device could then be triggered by a tank, or a person, or remotely
via a pullwire. In such circumstances the scale of civilian suffering has
been terrible. Some 6,000 amputations were necessary in 1990, with at
least double that number of people dying as a result of landmine explosions. One telling statistic is that one person in 256 is an amputee in
Cambodia whereas in the United States the number is one in 22,000.

Campaigns Against Mines

Given its traditional role in the alleviation of suffering caused by warfare,
the International Committee of the Red Cross has been strongly influenced by these recent uses of landmines. In a 1994 report it was noted
that an ICRC wound database instituted in mid-1990 held data on 17,414
patients from five independent ICRC hospitals dealing with injuries
from the wars in Afghanistan, Cambodia and Sudan.[11] Its concern led the
ICRC to convene a major symposium in Montreux in 1993[12] covering all
aspects of the landmines problem (Table 4.1). The symposium, in fact,
launched a worldwide campaign by Red Cross and Red Crescent national
societies to highlight the carnage being caused by landmines. Then, in
February 1994, the ICRC took the quite unprecedented step of calling for
a global effort to achieve a total ban on the use of anti-personnel mines.[13]

In its campaigning efforts the ICRC has been joined by a network of
NGOs around the world such as Human Rights Watch, Vietnam
Veterans of America Foundation, Handicap International and many others. Obviously, groups interested in conflict prevention, and medical
practitioners concerned with helping victims, were prominent in the
campaigns.[11] However, given the social and economic consequences of
the vast use of landmines, other NGOs concerned with development
issues have also become heavily involved. In Britain, for example, the
highly-respected development aid charity Oxfam carried a report,
'Hidden Killers', in its newsletter for supporters in the summer of
1994.[14] The article began with a true story:

> A victim of a mine lost a leg. He was fitted with an artificial limb and
> went back to work in his fields. He lost the other leg trying to rescue
> a friend who had fallen victim to another mine. Then, when yet
> another person set off a mine, he was blinded . . .

Table 4.1 Subjects investigated at the Montreux symposium on anti-personnel mines*

---

1. The present use of mines
    - by government armed forces;
    - by other armed groups.

2. The trade in mines
    - manufacturers, exporters, importers;
    - types of profits involved.

3. The humanitarian consequences
    - medical effects;
    - extent of the casualties;
    - social consequences;
    - rehabilitation needs versus resources available.

4. Technical characteristics of anti-personnel mines
    - self-destruct or neutralising mechanisms;
    - methods for rendering mines detectable.

5. De-mining
    - technical methods;
    - organisational necessities.

6. Professional military perceptions
    - of the use of landmines;
    - of the possibilities of recording;
    - appropriate self-destruction or neutralisation methods;
    - uses of undetectable mines.

7. The legal situation
    - presently applicable law;
    - explanations for limited participation in the IWC;
    - explanation of the limited legal provisions.

---

* From reference 12

The author commented that the man's tragic history may be extreme, but he is just one of the thousands of victims of landmines around the world every month. Oxfam is responding to the landmine disaster by helping to pay for de-mining operations in Africa and by assisting with the rehabilitation of victims in Cambodia. Crucially, also, as the article explained, 'we are taking action to prevent things getting worse – by calling for a worldwide ban on the use, production, stockpiling, and sale of APMs'. The co-operative effort was made clear:

. . . In co-operation with other agencies whose work is affected by mines, we are putting the case for this ban – especially to meetings at which changes to the UN's Inhumane Weapons Convention are being discussed.

This clear linkage of the devastation caused by warfare to the damage it does to prospects for development, and therefore the need to take preventive arms control action, is comparatively new and perhaps gives hope for much more effective action in the future.

Another striking aspect of the campaign against landmines has been the work of agencies concerned with the welfare of children. As a 1994 publication by UNICEF, speaking of landmines, explained:[15]

The use of these deadly devices where they pose a threat to children clearly violates the Convention on the Rights of the Child, including, of course, the provisions under which states are obliged to protect children in armed conflict . . .

This text appears opposite a heart-rending picture of a father with his tiny child, the latter learning to use crutches at the ICRC hospital for war victims in Kabul. The executive director of UNICEF was also in no doubt as to what needed to be done. His Foreword ended:

. . . I would like to add my voice to that of the International Committee of the Red Cross and urge the international community to go one critical step further and adopt a total ban on the production, use, stockpiling, sale and export of anti-personnel land-mines . . .

The strength of the campaign was symbolised, perhaps, by an article by Boutros Boutros-Ghali in the journal *Foreign Affairs* in the autumn of 1994.[16] After reviewing the scale of the problem, and the efforts that have been made to deal with it, the UN Secretary-General turned to the question of what further now needs to be done. He was much in favour of strengthening the IWC and slowing proliferation through the imposition of bans on exports by manufacturing countries (such as that initiated by the US in October 1992 and extended for three more years in 1993). Boutros-Ghali was also in favour of efforts to stigmatise landmines as unacceptable weapons in the same category, in world opinion, as chemical and biological weapons, but he argued:

. . . ultimately, and urgently, the world needs to establish an international convention on mines. Its purpose should be to reach agreement on a total ban on the production, stockpiling, trade, and use of mines and their components . . .

He thought, much like the ICRC, that this would be the only effective way to deal with the extent of the killing, maiming and societal destruction being caused by landmines.

A report in the technical press in early 1995, on the last of the preliminary meetings in the run-up to the autumn review of the IWC, gave some sense of the forces involved in determining the progress of the anti-landmines campaign.[17] Thirty-one of the 42 states party to the Convention, 26 non-party states, the ICRC and several other NGOs met in January. It was agreed that restrictions in the Convention should be extended to internal armed conflicts, in addition to those between states, but progress in other areas was less clear-cut. Some parties agreed with the ICRC's efforts to achieve a total ban on landmines; others argued for a ban on mines without self-destruct mechanisms. According to the report, however, most attention was centred on a proposal that anti-personnel mines used outside marked, guarded and fenced minefields should have self-destruct mechanisms and should be detectable (in order that clearance be possible). This last position, significantly, was supported by France, Germany, the UK and the US.

The difficulties in achieving agreement were stressed in the report: for example, the reluctance of some states (led by China) to specify a minimum metallic content (that would allow detection); the objections of significant Third World states to provisions limiting exports; and the refusal of some states to consider verification or enforcement measures. In short, it was not at all surprising that humanitarian concerns did not triumph over military and other necessities at the eventual review conference.

Military Considerations

It is obvious from the foregoing discussion that very large numbers of anti-personnel and other types of mine are being produced around the world. While it is difficult to get a clear picture of the production and trade in these weapon systems, the careful survey carried out by Human Rights Watch and Physicians for Human Rights[4] suggested that China, Italy and the former Soviet Union have been the largest producers and exporters of standard anti-personnel mines in recent years. The survey pointed to a combined global production of between 5 and 10 million mines per year in recent decades, at a cost of some \$50 to \$200 million annually. Such figures are hard to establish precisely, of course, in part because there are vested interests with a desire to keep them away from the public eye. As Ferrucio Petracco explained to the Montreux symposium, landmines are classified as weapon systems and are therefore already covered by export regulations in most Western countries. Moreover, their rules are said to be tougher in regard to countries at war. Yet, paradoxically, huge quantities of these weapon systems do still reach war zones:[18]

> . . . Such being the case, one could draw the conclusion that either governments do not want to be kept openly informed or that they are misled by their own producers and/or brokers involved in such deals.

A further point which is clear from the work of Human Rights Watch is that although the value of trade in conventional landmines may be small in comparison with that of other weapon systems, it is only part of the story. For more sophisticated (remotely-delivered) landmines there is an additional, and valuable, trade in delivery systems. As the munitions themselves become more complex, the value of the weapon systems' trade will also tend to increase.[4]

This came over particularly clearly in a contribution from a British military expert at the ICRC symposium in 1993. Speaking on the basis of British practice, but at a sufficiently general level to cover parallel trends in other armed forces, Lieutenant Colonel N. H. Rollo first emphasised the importance of mines in military operations:[19]

> . . . The cost effectiveness of obstacles which include a combination of mines and [other] appropriate weapon systems, may be increased by up to a factor of three over what is achievable if [those] weapon systems alone were deployed. Thus in addition to their utility when deployed on their own, mines will continue to be required to enhance the effectiveness of other weapon systems across the spectrum of operations . . .

It is obvious that in such circumstances there would be resistance in professional military forces to a total ban on mines. In discussion, Lieutenant Colonel Rollo reiterated the point:

> . . . As barrier systems were rendered two to three times more effective if reinforced by anti-personnel mines, and as anti-personnel mines were cheap, they were certainly cost effective.

Since the military will resist giving up mines, what is really interesting is the military view of what the future of mines will be. Lieutenant Colonel Rollo saw a wide range of general and specific features which were likely to be developed (Table 4.2). These are related to the future of modern, high-speed-manoeuvre warfare, and to the development of military electronics. To see the evolution of landmines in its proper context, the evolution of anti-personnel weaponry as a whole since the end of the Second World War needs to be reviewed.

Table 4.2   Possible features of new landmines*

General features

- improved flexibility of use;
- improved speed of laying;
- reduced logistic, manpower and training needs;
- capability to lay mines over larger areas;
- adaptation for remote laying by a variety of systems;
- continued capability also to lay AP mines by hand.

Specific features

- self-destruction;
- switch on-switch off capability;
- anti-lift devices;
- programmable life;
- resistance to electronic countermeasures.

---

* From reference 19

The Evolution of Anti-Personnel Weaponry

Eric Prokosch, who, as we saw at the end of the last chapter, has long been involved in attempts to control this class of weaponry, has argued that the increased knowledge of the effects of conventional munitions gained from the Second World War lies at the root of current problems. These discoveries:[20]

> . . . suggested that there was great scope for improvement in the design of anti-personnel weapons. The size of fragments could be drastically reduced, to the point where they had no more than the 'optimum energy' needed to inflict a fatal or severe wound.

The key to improving the effectiveness of a whole range of weapons was therefore to control the process of fragmentation so that a very much larger number of smaller, but very dangerous pieces could be spread evenly over the widest possible area.

This technical discovery was, according to Prokosch, given extra impetus by the nature of the major wars in which the leading country – the United States – became involved. In Korea the US first encountered the problem of fighting against massed infantry which prompted the need to find means for a mechanised army to deal more effectively with attack by massed, lightly-armed forces. After the initial development of the weaponry, the United States became involved in Vietnam, where it faced an enemy dispersed in terrain where its forces could not be easily located. This led to the initial field testing of the new weaponry by the United States.

One example of this process was the development of aerial mines. Their origins can be traced back to a US Air Force study which began within a year of the outbreak of the Korean War. Three new air-delivered mines were used in Vietnam: the small gravel or 'teabag' mine, which could be laid from helicopters and aeroplanes as well as by individuals or from a truck; the 'dragontooth' tiny plastic mine; and the wide-area, anti-personnel mine which sent out eight tripwires when it hit the ground and was therefore called a 'spider' mine by the Vietnamese. Huge numbers of these mines were used in Vietnam, Prokosch suggesting that the US Defense Department ordered over 37 million of the gravel mines alone in Fiscal Years 1967 and 1968. The same process of technical evolution in munitions may be seen in the development of cluster bombs, which consist of numbers of small bomblets which individually explode into small fragments after being spread over a wide area.

There was a major problem with US achievements in the technology of anti-personnel weaponry: they could not keep the military advantage gained for themselves. As Prokosch explained:

> . . . once a weapon is used, or sold to other countries, the secret is lost. The advantage is diffused as more nations acquire the weapon or make use of the surrounding technologies . . . the overall result is an increase in the efficiency and destructiveness worldwide.

So the technology of cluster bombs has evolved and spread, as has that of anti-personnel mines. The Soviet use of anti-personnel mines in Afghanistan in many ways mirrored that by US forces in Vietnam. Furthermore, as we have seen, forces with less sophisticated capabilities have subsequently found means to use the new military technology for their own purposes.

An important illustration of the general dangers of anti-personnel weapons proliferation arose in mid-1995. The ICRC's concern over the need to control blinding lasers had caused them to issue a campaign brochure in the autumn of 1994[21] entitled 'Blinding weapons: Gas 1918 . . . Lasers 1990s?'. The front cover of the brochure had a harrowing picture from the First World War of a line of soldiers with covered eyes walking with hands on one another's shoulders, in order to find their way. Yet, despite the obvious dangers, Human Rights Watch was able to report in May 1995 that:[22]

> The United States has pursued the development of at least ten different tactical laser weapons that have the potential of blinding individuals . . .

Moreover, while the Swedish government had suggested that a new protocol be added to the IWC in order to ban such weapons, the US Army, according to Human Rights Watch:

> . . . is hoping for a government decision next month [June 1995] to approve start of a full-scale production contract for the portable Laser Countermeasure System (LCMS), which is mounted on an M-16 rifle . . .

This time, however, it appears that the US was behind the game. The technical press in mid-1995 was already reporting that:[23]

> Norinco of China has begun openly to solicit sales enquiries for a device that it designates the ZM-87 portable laser 'disturber', an eye-damage weapon of a kind that the International Red Cross Committee has been campaigning to have banned from international sale . . .

At one level it might be argued that this is just another single weapon system whose regulation will need to be addressed before too long. Yet the import of Prokosch's argument is that we have to stop seeing each of these weapon developments in isolation. Instead, we must consider the whole range of anti-personnel weapons and their evolution in a proper historical context. What has happened with small calibre bullets, fragmentation grenades, cluster bombs, fuel-air explosives, incendiaries, remotely-delivered mines, hand-held lasers and other more exotic weapons may then be fully understood and a more appropriate approach to control attempted.

With the necessity of such an integrative perspective in mind, we begin in the next chapter to examine another kind of anti-personnel system which was intensively used by US forces in Vietnam – non-lethal chemical weapons. Though their history is largely forgotten today, it forms an essential context for understanding the more recent development of such weapons that will be described in later chapters. Prokosch succinctly summarised the reason why the IWC had failed to control anti-personnel mines:[20]

The problem of unexploded mines is an eloquent testimony to the failure of the efforts in the 1970s to adopt new bans on especially injurious and indiscriminate weapons . . . in 1974 Sweden and six other countries had proposed an outright ban on the aerial emplacement of anti-personnel mines. In the ensuing discussions, military considerations won the day, and the resulting Protocol II . . . is full of loopholes . . .

.5.

LETHAL AND NON-LETHAL
CHEMICAL AGENTS

Any consideration of chemical weapons has to begin with the effective deployment and use of these weapons in the First World War.[1] Discoveries in chemistry during the 19th century, and the development of chemical industries in the major industrial powers, made available large quantities of agents such as chlorine and phosgene. The dangers of using asphyxiating gases were recognised and The Hague Conventions, agreed at the turn of the century, banned the use of projectiles whose only purpose lay in the diffusion of asphyxiating or deleterious gases. Unfortunately, the prohibition against the use of chemical weapons did not last in the terrible conditions of the Great War.

Both sides in the war recognised the possibility of gas warfare and made some preparations which allowed the use of non-lethal agents such as tear gas very early. We shall return later to the issue of non-lethal agents but first we need to consider lethal chemical warfare. At Ypres in the spring of 1915 the German forces first attacked the French with chlorine gas released from cylinders and allowed to drift over the opposition's trenches on the wind. This had a devastating effect on the French forces, but the attack was not followed up. Later chlorine attacks on British and Canadian forces were less effective since the troops were not caught totally unprepared. The exact number of casualties from these first uses of lethal chemical weapons is unknown, but the process which would lead to over a million casualties and some 90,000 deaths from gas attack during the war was underway.

Although during 1916 and the first half of 1917 new gas agents such as phosgene were introduced by both sides, and various forms of artillery shell and mortar were brought in to replace cylinders as the means of delivery, there was no significant breakthrough in terms of military effectiveness. Then, in mid-1917, the Germans introduced mustard gas shells. These were first used against the British at Ypres in July. Within two weeks over 10,000 men had been treated in British casualty clearing stations as a result of the use of the new agent. The mortality rate among mustard gas casualties was lower than for other lethal gases but it affected the skin directly, as well as the lungs, and its long-term effects could be very debilitating. The initial effects were appalling:[1]

> . . . Initially it irritates the eyes and throat, intensifying after some hours to intense pain in the eyes, with blisters developing on any skin that has come into contact with the agent, which can penetrate through several layers of clothing. Death results from the damage to the respiratory system . . .

At a military level though, gas warfare was usually not particularly effective. This was partly due to the novelty of the new weaponry for those attempting to use it, but mainly because of the rapid introduction of defensive countermeasures, such as protective masks, on both sides. This led to a search for means of adding other agents which might penetrate masks and disrupt the protective measures, but the war ended with no really significant developments being achieved.

During the Second World War both sides held large stocks of chemical weapons but, mercifully, these were not used. Various forms of mustard gas were also still held by the major powers when the Chemical Weapons Convention was agreed at the end of the Cold War. However, the major innovation after the First World War resulted from the discovery of the much more lethal nerve gases in Germany between the two World Wars. These formed the key element in the chemical arsenals of the major powers during the Cold War period. Both mustard gas and nerve gases were used on a militarily significant scale during the long Iran–Iraq war of the 1980s.

Lethal Chemical Agents

This brief overview is an attempt to summarise the characteristics of the main types of lethal chemical weapon agent[2] that have been developed and deployed this century (Table 5.1).

Choking agents cause physical injury to the respiratory tract.[3] Phosgene, an example of a choking agent, has had many industrial applications and has been produced in large quantities in many

Table 5.1 Categories of lethal chemical weapons*

Type	Examples	Concentration to kill half those exposed for 1 minute (milligrams per cubic metre)
Choking agents	Chlorine	19,000
	Phosgene	3,200
Blood agent	Hydrogen cyanide	1,000–5,000
Blister agents	Mustard	1,500
	Nitrogen mustard	> 1,500
	Lewisite	1,200–1,300
Nerve agents	Tabun	150–400
	Sarin	100
	Soman	50–100
	VX	10–50

* From reference 2

countries. Except in cold weather, it is a colourless gas which is denser than air and thus sinks towards the ground. When inhaled by the victim its initial effects may be too mild to warn of danger and the symptoms of the damage in progress are generally not obvious for some hours following inhalation. Severe exposure can produce convulsions and death. A large proportion of the deaths resulting from chemical agents in the First World War were caused by phosgene and stocks were held by both sides during the Second World War.

The **blood** agent hydrogen cyanide was much more difficult to use effectively since, being less dense than air, it was not possible to build up lethal concentrations of the gas near the ground. Blood agents are so called because they interfere with oxygen transport in the blood.[2] Hydrogen cyanide is readily absorbed through the lungs and death occurs very rapidly if lethal concentrations are encountered.

The causes of the effects of the **blister** agent, mustard gas, on any exposed human tissue have been the subject of extensive research since the First World War.[4] They are clearly very complex and involve, for example, disruption of cellular functions through binding with DNA (the genetic material of the cell). The characteristic skin blisters caused by exposure to mustard gas are slow to heal and secondary infections can easily occur. Additionally, longer-term consequences, such as cancer of the respiratory tract, are common.[5]

By far the most important lethal chemical weapon agents at present are the **nerve gases**. As can be seen from Table 5.1, a far smaller amount of these chemicals is required to kill a person than for other agents. These poisons can be absorbed through the lungs or direct through the skin. To understand their mode of action more readily, it is necessary to outline first how a nerve cell – or neuron – functions within the nervous system of humans. Such a description of the individual unit of the human nervous system will also be helpful in laying foundations for the discussion of the nervous system as a whole, and its disruption by non-lethal chemical agents, in later chapters.

The neuron and synaptic transmission

The human brain is thought to contain some 100 billion individual neurons.[6] That the system is made up of these individual cellular units only became clear at the start of this century with the discovery by Camillo Golgi of a vital stain which coloured completely just an occasional cell in any one physiological preparation of nervous tissue, and the systematic application of this staining technique by Ramón y Cajal.

Neurons take many different forms but each one basically consists of a cell body which contains, within a central nucleus, the genetic material controlling cellular functions and a number of projections (Figure 5.1 at A). The role of the neuron is to transmit information. This it does by sending discrete electrical signals, termed **action potentials**, down one generally longer projection called the **axon**. The other projections are called **dendrites** and, in the central nervous system (CNS), their function is to receive information from other nerve cells. We are primarily concerned here with the question of how information travelling down the axon of one neuron is passed on to another neuron, or to a cell in a neuro-effector organ such as a muscle fibre in the skeletal muscle of an arm or leg.

The axon typically divides into a set of branches as it approaches another nerve cell or a muscle fibre and high-power electron microscopic images show that these branches form typical structures, called **synapses**, with the neuron or muscle fibre (Figure 5.1 at B). Similar structures are formed on the dendrites of the first neuron by axon branches from other cells. Earlier this century Otto Loewi discovered that a nerve cell could communicate with other cells by means of a chemical messenger. To demonstrate this, Loewi used a physiological preparation of frog hearts with saline solution (perfusion fluid) flowing

Figure 5.1   A Generalised Neuron (A) and Synapse (B)*

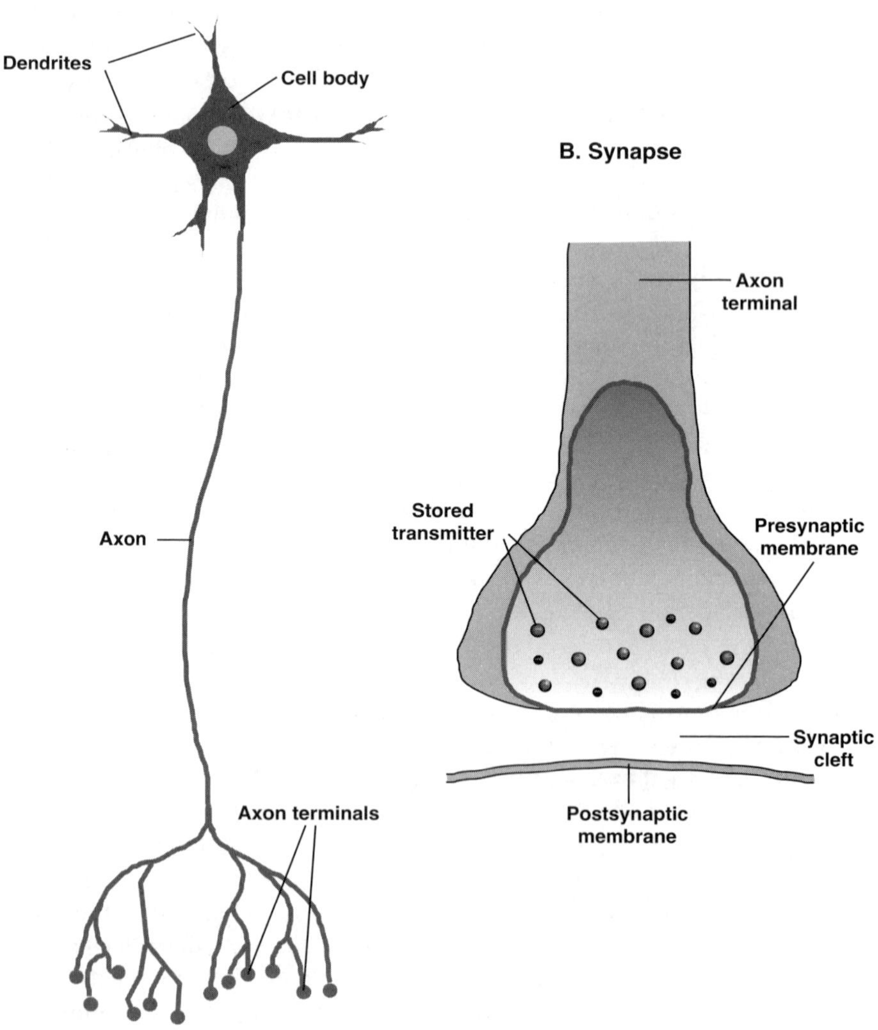

* After reference 6

through them. He electrically stimulated the vagus nerve to one heart, thereby slowing (inhibiting) its rate of muscle contraction or beat. He then passed the perfusion fluid from the first heart into a second heart whose beat rate similarly slowed but in the absence of stimulation of its vagus nerve. Loewi concluded that when the vagus nerve to the first heart was stimulated electrically a chemical substance was released which slowed the rate of its muscle contraction and that this substance then acted in a similar way on the second heart.[7] The process of chemical communication is called **neurochemical transmission** or neurotransmission and the chemicals released are called **neurotransmitters**.

The process of transmission takes place at the synapses between the terminals of the axon branches and the **post-synaptic** cell – either another neuron (a neuro-neuronal junction) or some other cell type such as a muscle fibre (a neuromuscular junction). The neurotransmitter substance is stored in vesicles in the pre-synaptic axon branch terminal, and is released in small amounts when an axon potential arrives at this terminal. Clearly, the more axon potentials that arrive, the more neurotransmitter that is released. So if a **pre-synaptic** neuron is very active it will have a stronger effect on post-synaptic cells than if it is comparatively quiescent. The signalling system can be very complex. Because a transmitter exerts its effect when it crosses the **synaptic cleft** between two cells and attaches to appropriate **receptors** on the post-synaptic cell, the same transmitter can have quite different effects – either excitation or inhibition – on different post-synaptic cells, depending on the receptor type each possesses. There necessarily has to be some mechanism for deactivating the transmitter once it is released into the synaptic cleft or its effects will persist long after the action potential which caused its release has occurred.

The substance which Loewi discovered to slow the heart in frogs has important functions in the human nervous system. It is called **acetylcholine** (ACh) and it is the transmitter both for the excitatory motor neurons which control the skeletal muscles of our bodies and for some parts of the autonomic nervous systems which regulate our internal organs where, for example, it acts to slow the heart. As we shall see, ACh is also a transmitter in some parts of the brain. ACh is manufactured in the cell and is stored in vesicles at the axon terminals. When it is released from a terminal into the synaptic cleft, on arrival of an action potential or electrical stimulus, it attaches to the post-synaptic receptors and affects the electrical activity of the post-synaptic cell. However, it is also very rapidly broken down by an enzyme called **acetyl-**

cholinesterase (AChE) and the constituent parts of the transmitter molecule are then taken back up into the axon terminal and reused to make new transmitter.[6]

When ACh is the transmitter substance the synaptic system is termed **cholinergic**. Drugs and poisons can affect the cholinergic neurotransmission system in many different ways. For example, botulinum toxin, which is sometimes found in poorly canned foods, prevents the release of ACh from presynaptic terminals. Such sensitivity to chemicals has provided a means for distinguishing between two different types of post-synaptic receptor for ACh. Nicotine, the biologically active ingredient in tobacco, mimics the effect of ACh at synapses on skeletal muscle but has no effect on heart muscle cells. On the other hand, muscarine, a mushroom extract used by humans as an hallucinogen, has no effect on skeletal muscles but mimics the effect of ACh on the ACh synapses of heart muscle cells. The first type of ACh synapse is said to be **nicotinic** and the second to be **muscarinic**. This kind of information is helpful in understanding the effects of biologically active chemicals such as curare, which causes paralysis because it blocks the nicotinic receptors of the neuromuscular junctions in human skeletal muscles.

Such complex issues as the sub-types of synapses will be discussed again later. Here, with the basic knowledge of neurons and synaptic transmission, we turn to the question of how modern lethal nerve agents function.

The action of lethal nerve agents

Nerve agents were first synthesised in Germany from the mid-1930s. Tabun was the first agent discovered, then sarin and finally, in 1944, soman. By the end of the Second World War, 12,000 tons of tabun were stockpiled in Germany as well as some 600 tons of sarin. These agents are usually referred to as the G series: GA (tabun), GB (sarin) and GD (soman). The even more toxic V agents were developed during the 1950s. All these agents are organophosphorous anticholinesterases (anti-AChEs), which exert their effect by inhibiting the action of the enzyme acetylcholinesterase, which normally breaks down ACh after use at its cholinergic synapses.[4] The result is that ACh accumulates in the synaptic cleft and its normal effect is disrupted by the continuous stimulation exerted. A significant feature of these nerve agents is the very small amount needed for their effects to be manifested. Given the known locations of ACh synapses in the skeletal muscles, the autonomic nervous system and the brain, it is hardly surprising that nerve agents exert a very wide range of effects on bodily functions. Death is usually caused by oxygen deprivation resulting from respiratory failure.

Knowledge of the way that nerve agents act on ACh synapses has allowed the development of a range of potential palliatives. Soldiers can be pre-treated with drugs which bind reversibly to AChE. This protects the enzyme during exposure to nerve agent; following exposure, the demand for the enzyme leads to its release from the reversible binding. Another approach is to allow soldiers to self-administer a drug such as atropine (belladonna), which blocks some ACh receptors. Though on exposure to the nerve agent this mitigates the effects of the excess of ACh, because atropine is a powerful drug in its own right, auto-injection by mistake – say in panic over possible nerve gas attack – can itself incapacitate. Finally, drugs have been discovered which break the bond between the AChE enzyme and the nerve agent. These so-called oxime drugs are considered to be a major step forward in treatment.[4]

Having briefly introduced the subject of anti-personnel chemical weapons and seen something of the way in which a knowledge of the nervous system in humans can be used and misused, we now return to the story of the evolution of anti-personnel weapons. In particular, we shall look at the use of non-lethal chemical agents by the United States during the Vietnam War. To do so, it is necessary to discover what happened to chemical weapons control between the end of the First World War and the Vietnam War, and to see how the US viewed these developments.

Non-Lethal Chemical Agents

At the time of the Vietnam War, the Stockholm International Peace Research Institute (SIPRI) produced a series of six volumes on *The Problem of Chemical and Biological Warfare*. There will be cause to refer to a number of these volumes. They were produced collectively by the institute, but where it is possible to identify the authors of individual volumes this will be done.

In the introductory volume[8] SIPRI suggested that there were two broad categories of agent: casualty and harassing. A casualty agent was defined as 'a CW agent whose normal field concentrations are capable of causing severe injury or death'. These agents were intended to put enemy forces out of action for a long time or to cause death, and we shall therefore refer to them here as **lethal** agents. The second category, the harassing agent, was defined as 'a CW agent whose normal field concentrations are capable of rapidly causing a temporary disablement that lasts for a period not greatly exceeding that of exposure'. The main function of harassing agents was not directly to kill or severely injure opposition forces but to 'force the enemy to put on respirators or to disconcert his combat activities'. Two main types of such **non-lethal** agent were developed during the First World War. These were agents producing lachrymation – a flow of tears – and those producing sternutation or sneezing. Examples of tear gas developed during the war were α-bromobenzyl cyanide (CA) and α-chloroacetophenone (CN). Examples of sternutators were diphenylchloroarsine (DA), diphenylcyanoarsine (DC) and adamsite (DM).

A NATO document of the early 1970s added to this classification in two ways (Table 5.2). First, another group, 'Other Sensory Irritants', was added to the lachrymators and sternutators. Specifically:[9]

> . . . CS is the code-name for orthochloro-benzylidene malononitrile. On account of its stronger irritant effect, its wider range of attack (eyes, respiratory tract, and skin) and its lower toxicity it has fully superseded CN . . .

The fact that CS equally affected the eyes and upper respiratory tract led to its separate categorisation. Secondly, an entirely new category was added – 'Incapacitants'. An incapacitant chemical:

> . . . produces a temporary disabling condition that persists for hours to days after exposure to the agent has ceased (unlike that produced by riot control agents) and medical treatment, while not required, facilitates a more rapid recovery . . .

The text went on to explain that these agents were highly potent, produced their effects by disruption of higher functions of the central nervous system (CNS), had an effective duration of hours or days, but did not endanger life at the dosages used or cause permanent injury. It was suggested that only two types of incapacitant were of military interest: CNS depressants such as the agent BZ and CNS stimulants such as lysergic acid diethylamide (LSD).

As the 1970s SIPRI study of chemical weapons pointed out,[10] in the early post-Second World War days the US Army Chemical Corps was interested in a very wide range of possible means of incapacitation. These included causing fainting through the sudden lowering of blood pressure by the use of anti-hypertensive drugs. Another approach was to find new means of causing retching and vomiting, or of disturbing body temperature. The crucial problem was to find chemicals both effective at the kinds of low dose at which nerve agents worked and with a very

Table 5.2   Categories of non-lethal chemical weapons *

Type	Examples	Substances
Harassing/riot control agents	Lachrymators (tear-gas)	CA CN
	Sternutators (sneezing agents)	DA DC DM
	Other sensory irritants	CS
Incapacitants	CNS-depressants	BZ
	CNS-stimulants	LSD

* From reference 9

wide gap between an effective incapacitating dose and a lethal dose. The second of these requirements was difficult to meet and thus the range of dosages likely to be encountered in an effective field application would have included some in the lethal range. We shall consider these issues again in later chapters and now discuss further the military uses of riot control agents.

The 1925 Geneva Protocol

Following the peace treaty, the first multilateral attempt to deal with the problem of chemical weapons after the First World War was made in the Washington Treaty of 1922. This was concerned with limiting the use of submarines and prohibiting the use of chemical weapons in war. Although the Treaty was ratified by the United States, entry into force required unanimous acceptance by all those concerned. This did not happen because France objected to the clauses dealing with sub-marines.[11] However, the text of the Washington Treaty clearly formed the basis of the 1925 Geneva Protocol. The use of chemical weapons in war was considered at a conference on international trade in armaments as a result of US initiatives and was formulated as a protocol, since it was not directly related to the formal purpose of the conference.[12] The protocol states, in part, that:

> Whereas the use in war of asphyxiating, poisonous or other gases, and of all analogous liquids, materials or devices, has been justly con-demned by the general opinion of the civilised world . . .

and therefore that the states party to the protocol:

> Declare: That the High Contracting Parties, so far as they are not already Parties to Treaties prohibiting such use, accept this prohibition, agree to extend this prohibition to the use of bacteriological methods of warfare and agree to be bound as between themselves according to the terms of this declaration.

Unfortunately, reservations entered by various states reduced the prohibition essentially to a no-first-use agreement between the parties. Worse still, lobbying by supporters of chemical warfare prevented ratification of the protocol at that time by the United States.[11] Nevertheless, and particularly because President Roosevelt was a very strong opponent of chemical warfare, during the Second World War the United States made it clear that it too would use such weapons only if they were used first by the other side.

While it might therefore be argued that, although the US was not a signatory to the protocol it was nevertheless bound by customary law to observe the convention among civilised states, there arose a more immediate question of interpretation during the Vietnam War. The US government then argued that the prohibition in the protocol did not apply to the use of tear gas (or chemicals such as Agent Orange which affected plants). This view did not meet with the agreement of many other states. It seems to have been based on a possible difference between the French and English texts but, as a 1970 article in the *International Review of the Red Cross* noted:[13]

> . . . the French Government, the depository of the Geneva Protocol and the first to ratify it, specified in a note in connection with the preparatory work leading to the League of Nations disarmament conference, that it considered the prohibition extended to the use of lachrymatory agents . . .

The arms control expert, Joseph Goldblat, who wrote the detailed SIPRI study on this question, summarised his views in an article in the *Bulletin of the Atomic Scientists*.[14] He pointed out that in the discussions on improved disarmament measures in the League of Nations during the early 1930s, the British had demonstrated that this was their opinion also. He quoted a UK draft convention submitted to the General Commission of the Disarmament Conference of 1933 which stated:

> . . . The prohibition of the use of chemical weapons shall apply to the use, by any method whatsoever, for the purpose of injuring an adversary, of any natural or synthetic substance harmful to the human

or animal organism, whether solid, liquid or gaseous, such as toxic, asphyxiating, lachrymatory, irritant or vesicant substances.

Goldblat also emphasised that during the discussions in the 1930s the United States had not made any significant objection to this interpretation of international law. The US was mainly concerned with ensuring that the option of using tear gas was retained for domestic police forces. This option is obviously legally retained by signatories to the Geneva Protocol of 1925 because it only applies to the use of chemical agents 'in war'.

Domestic use of tear gas

Towards the end of the period of research on potential harassing agents during the First World War a lachrymatory agent, α-chloroacetophenone (CN), was produced, which was adopted by police forces in the United States after the war.[15] According to SIPRI's account:[8]

> . . . Exposure to a CN aerosol concentration exceeding about 0.5 mg/m^3 induces a copious flow of tears in less than a minute. At higher concentrations . . . intense irritation is experienced in the nose and upper respiratory tract, soon followed by an itching and burning of moist areas of exposed skin, which may even lead to blistering . . .

Though recovery after exposure is usually swift, the account noted that lung damage might occur and that a number of deaths through pulmonary oedema had been reported. The attraction of this chemical for police force use was pronounced. Colonel Rex Applegate, a prominent American military author of the Vietnam period, noted:[15]

> . . . Police CN usage increased during the 1930s . . . .The CN agent was disseminated from various devices, utilising the burning or liquid tear-gas method of dissemination . . .

During the Second World War further research produced 'micro-pulverized' CN, whose talc-sized particles were used to produce longer-lasting and greater irritation.

Applegate's manual lists a wide range of situations in which domestic police might use tear gas: for riot control and mob dispersion; against a barricaded fugitive; for individual protection; in prisons and mental institutions; and in the protection of banks, vaults and safes. He suggested that CN was the preferred type of tear gas among law enforcement officers and expressed reservations about the newer CS gas:

> . . . During the disturbances in Washington, DC in April, 1968, many stores received heavy concentrations of the CS dust-type agent. In

some cases, and in spite of efforts to decontaminate, the premises could not thereafter be occupied for several weeks.

Despite this, chemical agents were held in high regard on the whole. In 1972 a US National Science Foundation study, *Non-lethal Weapons for Law Enforcement*, concluded that, apart from the traditional night-stick, chemical agents were the only such non-lethal weapon in effective and widespread use.[16] A follow-up study in 1987 concluded that:[17]

> . . . Tear gas has been standard in police inventories since the late 1960s. Officers frequently carry personal-issue hand dispensers, and most departments have tear gas shells for shooting dispensers past barricades. Large-volume dispensers can be used for crowd control.

Moreover, the effectiveness of CS as compared with CN had by that time led to its replacing the older agent in the UK and US.

Originally synthesised in the United States in the 1920s, CS was developed in the UK in the 1950s.[18] Official investigations of its effects following large-scale use in Londonderry in 1969 suggested that its possible ill-effects after civil use were relatively slight,[19] but doubts persist in medical circles that this is so. One report in a major American medical journal in 1989 argued:[20]

> Inhalation toxicology studies at high levels of CS exposure . . . have demonstrated its ability to cause chemical pneumonitis and fatal pulmonary edema. In situations in which high levels of exposures have occurred, the same effects, as well as heart failure, hepatocellular damage, and death, have been reported in adults . . .

Our major concern here, though, is with the use of non-lethal chemical agents in warfare. There will clearly be some situations where military forces may use tear gas in an analogous manner to civil police, for example, in containing a distubance within a prisoner-of-war camp. However, the military use of tear gas as an adjunct to firepower in warfare is a very different matter, as the US practice in Vietnam demonstrated.

The Use of CS Gas in Vietnam

According to Meselson's account in *Scientific American* in May 1970, the United States did not at first intend to go beyond riot control situations in its use of tear gas in Vietnam. He quoted Secretary of State

Dean Rusk's statement of 24 March 1965, following early press reports of tear gas use, that:[21]

> We do not expect that gas will be used in ordinary military operations . . . .The anticipation is, of course, that these weapons will be used only in those situations involving riot control or situations analogous to riot control.

But five months later a marine apparently used CS grenades to remove 400 civilians and some enemy soldiers from a cave, without injury to the civilians. Following this demonstration of the 'humanitarian' use of CS to deal with a situation where combatants and non-combatants were intermingled, much wider use of CS followed. SIPRI's account noted that by the beginning of 1966 irritants were being widely used in regular combat operations.[8] Indeed, while the SIPRI account accepted that the small amounts of CS initially supplied to US forces were probably intended for use in the 'intermingled' situation, it noted:

> . . . it became known at the end of 1969 that the authorisation given in the autumn of 1965 for continued use of CN and CS in fact placed no restrictions on the manner in which these agents were to be used . . .

Specifically, it continued:

> . . . There was no directive that they should be used only for humanitarian purposes. They were to be regarded as a normal component of the available inventory of weapons . . .

With vested interests supporting the use of chemicals, and a rapidly escalating war, this appears to have opened the floodgates to an exploration of what could be done with the weapon system.

The initial stages began in 1966:[8]

> Although CS continued in frequent use throughout 1966 and 1967, it was still largely an experimental or special-purpose weapon. Field commanders had not yet appreciated all its possible applications . . .

Yet in a process which seems to mirror that discussed in relation to other conventional weapons in the last chapter, SIPRI argued that:

> . . . A whole range of new CS munitions were being developed in the USA and sent out to Viet-Nam [*sic*] for testing, but it took time for appropriate employment doctrine to evolve . . .

A detailed account of the use of CS in Vietnam pointed to one of the mechanisms involved.[22] The relevant US Army Training Circular (3-16) on *Employment of Riot Control Agents . . . in Counterguerrilla Operations* (issued in April 1969),[23] in its section on the employment of riot control agent munitions and equipment, urged users 'to submit after-action reports on all experimental munitions'. The result was reported by SIPRI to be that:

> . . . During 1968 nearly twice as much CS was consumed as in all the previous years put together, and in 1969 there was a further 20 per cent increase . . .

The amounts of CS used in operations in Vietnam became truly formidable[8,22,24] in the late 1960s (Table 5.3). However, as Meselson pointed out at the time,[21] a standard military counter-reaction process set in as the North Vietnamese troops were equipped with good masks which reduced the effects of CS, and therefore its utility to US forces.

Table 5.3 Annual procurements of CS (in pounds weight)*

Fiscal year	CS	CS–1	CS–2
1964	233,000	142,000	
1965	93,000	182,000	
1966	373,000	1,217,000	
1967	437,000	770,000	
1968	714,000	3,249,000	288,000
1969	2,018,000	160,000	3,885,000
1970	–	354,000	1,830,000
TOTAL	3,868,000	6,074,000	6,003,000

GRAND TOTAL
OF ALL TYPES **15,945,000**

* From reference 22

During its period of heavy use, the CS agent was employed in three different forms. The original form was disseminated by thermal volatilisation in burning munitions. Improved technology produced a micro-pulverised form (CS-1) that could be packaged in bulk disseminators or in bursting munitions. Finally, CS-1 coated with silicone produced the waterproofed CS-2 form which persisted for several weeks in the field where it could then be stirred up by movement of people.[23] An extraordinary range of weapon systems was developed to disperse these CS agents, ranging from hand grenades through mortar and artillery shells to means of bulk dispersal from helicopters and high performance aircraft.[8,22] Some examples are given in Table 5.4. These weapon systems were then used in a variety of operations.

Table 5.4 Examples of CS munitions*

Weapon	Delivery system	Agent payload in kg.	Rate of CS discharge in grammes/sec.
M–7A3 hand grenade, CS	Hand or rifle	0.12	6
XM–630 cartridge, 4.2 inch, CS	Mortar	0.9	4, for each of 4 canisters
XM–631 projectile, 155 mm, CS	Howitzer	2.2	5, for each of 5 canisters
M–106 dispenser, CS-1 or CS–2	Portable	3.3 per hopper	200
M–4 bulk agent disperser, CS–1 or CS–2	Helicopter or vehicle	22 per hopper	1,100
CBU–30/A canister dispenser, CS	Aircraft	25	1, for each of 1280 canisters

* From reference 8

One major type of operation in which CS was used was the attempt to drive Vietnamese out of underground hide-outs. The M–7A3 hand grenade was often used for this purpose but larger quantities of agent could be introduced through use of the M–106 'Mity Mite' dispenser:[22]

> . . . It is a commercial agricultural dust-sprayer modified for military use. With its blowing capacity of 450 cfm of air, it can blow 10 pounds of CS or CS–1 into a shelter every 3–4 minutes . . .

Whilst there is said to be a very wide gap between the effective and the lethal dose for CS, it seems unlikely that much control over dosage levels was possible in this or other applications in Vietnam.[23]

A second major use of CS was its application over large areas of land in an attempt to render them uninhabitable for the enemy. CS–2 was obviously useful in this kind of operation because of its persistence. An idea of the magnitude of CS use can be gained from the SIPRI description:[8]

> . . . An early technique for spreading the CS, one which came into frequent use early in 1966, was to air-drop 55-gallon shipping containers of the agent that had been fitted up with improvised explosive bursters time-fused to detonate just above the ground. A CH–47 helicopter can dispense 30 such drums per sortie, each containing about 35 kg of CS . . .

Later, a standardised XM-28 bagged agent dispenser was developed for helicopter use. Another major use of CS was for exactly the same purpose that older harassing agents were used in the First World War, that is, to enhance the effects of conventional weaponry. Examples, as might be expected, include co-ordinated use of CS and artillery where casualties caused by the artillery were increased because people had been been driven into the open by the CS gas. Given the importance of helicopter operations in Vietnam, CS was also employed shortly before a landing in order to suppress enemy fire and force people out of cover.

All this military action came to naught and the United States eventually left Vietnam in defeat, but the use of harassing agents left unresolved problems. When President Nixon sent the Geneva Protocol to the Senate again for ratification he did so with a statement that:[11]

> . . . as a matter of US policy, the non-lethal agents were not included in the US definition of CW . . .

He did not make this a formal, legal reservation but this was still not acceptable to the Senate Foreign Relations Committee, so the protocol was not ratified for a further four years. The Ford administration eventually agreed a compromise that was acceptable to the Senate in which:[11]

> . . . the harassing agents would be limited to 'defensive military modes to save lives, such as controlling rioting prisoners of war, protecting civilians being used to mask or screen attacks, rescue of personnel such as downed air crews in remotely isolated areas, and protection of convoys in rear echelon areas' . . .

This compromise is still liable to cause problems for the future of the modern Chemical Weapons Convention, as we shall see in Chapter 9. The British also complicated the situation in 1970 by arguing, incorrectly, that CS use in war was consistent with the protocol on account of its low lethality.[25]

The most dangerous aspect of the US employment of CS in Vietnam was, however, the demonstration it gave that the most advanced military power in the world had a continuing interest in this type of weaponry. This was brilliantly explained to Congress in 1969 by Han Swyter who was, from 1966 to 1968, on the staff of the Assistant Secretary of Defense for Systems Analysis where he was directly concerned with chemical and biological warfare policies and operations. In a presentation significantly entitled 'Political considerations and analysis of military requirements of chemical and biological weapons', he emphasised the dangers of proliferation:[26]

> The Geneva Protocol is a major step in the direction of non-proliferation. It has provided the norm for international law and practice for 45 years. Tear gas and defoliants, as we are using them in Vietnam, tend to weaken the protocol. Our use enlarges a loophole. It tends to allow one kind of chemical and not another. Distinguishing among chemicals is not practical, no matter how good the intention . . .

Swyter reinforced his point by stressing that the potential attraction of chemical weapons would necessarily be greater for smaller and poorer nations. Thus proliferation of these weapons – as we have indeed seen since – would be easier than for nuclear weapons.

The question naturally arises as to why the United States started to use chemical agents in Vietnam in the way that it did. One factor of importance was obviously the growing strength of the Chemical Corps in the US Army. Clearly, the discovery of the new nerve gases by both Cold War antagonists when Germany was occupied in 1945 gave cause for reconsideration of the importance of chemical warfare.[27] There is ample evidence of Western concern over possible Soviet capabilities and intentions. One well-known Congressional Report, *Research in CBR (Chemical, Biological and Radiological)*, published in 1959, stated that:[28]

> The conclusion is inescapable that the Soviet Union and other Communist countries plan to use CBR if they find it to their advantage. Otherwise, their research effort would not have been continued to large-scale manufacture of materials and delivery devices. This country must make sure that it is not to their advantage to use this form of warfare against us.

In the SIPRI study of chemical and biological warfare written in the early 1970s it was suggested that during the late 1950s the Chemical Corps was engaged in a public campaign to increase its level of funding.[29] The opportunity arose in part from the move away from massive retaliation and towards a flexible response doctrine by the US forces as a whole, and the lifting of the prohibition on the first use of chemical and biological weapons by US forces in the mid-1950s. The proposition that chemical agents such as CS could be used in a 'humane' way formed part of the argument presented to Congress and the public, but how did CS come to be produced in the first place? To understand that, a broader perspective is required on thinking related to chemical weapons after the Second World War.

The Wider Context

The fiscal year report of the US Army Chemical Corps for 1959 relates how the British informed the US Army of its discovery of the utility of CS at the Thirteenth Tripartite Conference on Toxicological Warfare, held in Canada in 1958. The US Chemical Corps carried out its own evaluation and found the agent to be very effective. Subsequently, in the latter part of 1958, the Corps[30]

> . . . set up a crash program 'Black Magic' to produce the agent as a filling for tear and riot grenades, the Irritant Gas Disperser, the Helicopter Disperser, the Portable Gas Disperser, and the All-Purpose Army Aircraft Spray Tank. In June 1959 the Corps designated the compound as a standard agent . . .

The urgency of this operation is said to have originated from a concern that the Berlin situation might get out of hand. There, problems might have arisen in dealing with situations where combatants and non-combatants were intermingled. Shipments of CS munitions to the Panama Canal Zone are also mentioned.

According to a detailed historical monograph produced by the US Army Munitions Command, the Army did not begin to get interested in the battlefield use of CS until it became involved in Vietnam. The initial idea was to assist the landing of airborne troops, but a full-scale troop test code-named 'Water Bucket', carried out in June 1963, suggested that effectively dispersed CS could be used in a variety of situations. Soon the CS weapons were incorporated into the ENSURE (expedited non-standard urgent requirement for equipment) programme which accelerated development. As the monograph stated, confirming SIPRI's supposition:[31]

> . . . ENSURE was a method for putting weapons concepts desired by
> forces in Vietnam . . . through the engineering design phase as rapidly
> as possible and shipping limited quantities of the resulting munitions
> to Vietnam for field evaluation without waiting for completion of the
> standard research and development cycle . . .

Written in 1970, the report suggested that on the whole the process was
a success. It concluded that:

> . . . the prediction of the first Water Bucket test report in 1963 that a
> riot control agent could be a true tactical weapon if properly employed
> was borne out in the test of battle during the same decade . . .

Today, anxiety about the misuse of CS munitions by armed forces
remains widespread.[32] CS is nevertheless still widely, and legally, used
in an enormous variety of munitions by domestic police forces around
the world.[33] Another agent, **capsaicin** or 'pepper spray', is in widespread use among police forces in the United States.[34] In Britain,
however, the government has refused to allow its deployment, citing as
the reason unanswered concerns over its safety.[35] Since capsaicin could
be an indicator of what might lie ahead in any unrestricted evolution of
non-lethal chemical weaponry, we should know how it works.

Capsaicin: a sensory neurotoxin

To understand how capsaicin, a pungent and very powerful chemical
produced by pepper plants, exerts its action we need to look further at
the way in which the human nervous system is constructed. Earlier in
the chapter we discussed the structure and function of the individual unit
of the nervous system, the neuron, and the way in which it transmits
information electrically within itself and chemically to other cells. We
took as one illustration a **motor neuron** and its transmission of information to fibres of a muscle. This motor neuron and its muscle
connections may form part of a so-called **spinal reflex arc** whose other
components are a **sensory neuron** and an **interneuron**, two more of the
essential cellular components of the nervous system.

Sensory neurons have their sensory endings located at the periphery,
for example in the skin, and their cell bodies near the spinal cord.
Interneurons make up the vast majority of nerve cells in the brain and
spinal cord, that is, in the central nervous system, and may be connected to both sensory and motor neurons (as in the spinal reflex arc), or
just to other interneurons. The three neuron types, sensory, inter- and
motor, are illustrated in Figure 5.2, which shows a simplified cross-

section of the spinal cord. The cell body of the sensory neuron is located in the dorsal root ganglion of the spinal cord (paired ganglia or collections of nerve cell bodies are found in each body segment, a pair to each vertebra); the interneuron lies entirely within the spinal cord as does the cell body of the motor neuron whose axon runs out to the muscle. A sensory neuron receives *afferent* (*from* the periphery) stimulatory input, via the dorsal root, from its sensory receptor endings in the skin. The sensory neuron passes this information on, via its axon, to a connecting interneuron whose axon, in turn, is shown connecting with the motor neuron whose *efferent* output (travelling *to* the periphery) passes via *its* axon through the ventral root of the spinal cord to muscle. A so-called spinal reflex operates when a stimulus to the skin is relayed through such a circuit and causes a muscular response without direct involvement of the brain. To illustrate, a person may accidentally put a hand on a sharp object. Operation of the spinal reflex will mean that the hand is very rapidly withdrawn from the source of pain. Of course, there are also many spinal cord interneurons carrying the sensory information forward to the brain, and information originating in the brain can also affect the operation of the spinal cord cells.

Figure 5.2 Sensory, Motor and Interneurons in the Spinal Cord*

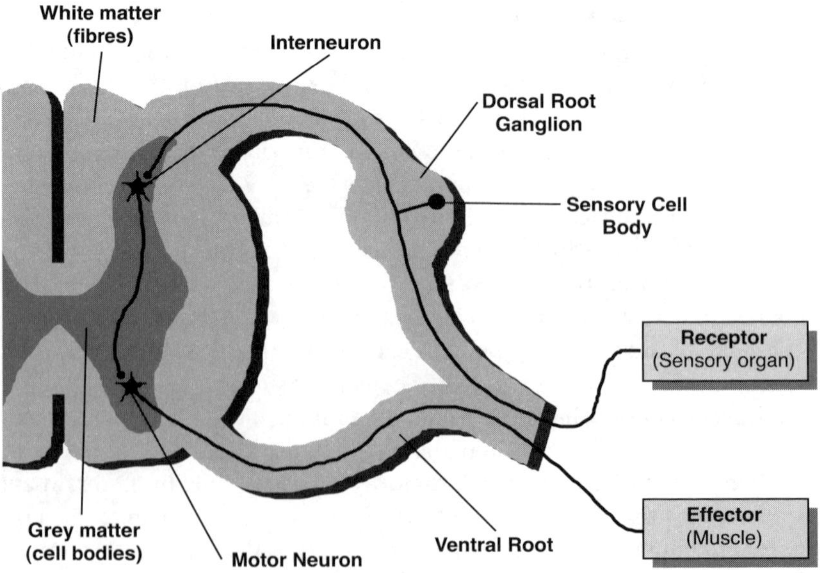

* After reference 6

A variety of types of sensory neuron innervate (supply) human skin. One of these cell types is the **C polymodal nociceptor**. Such cells are common in skin and respond to noxious mechanical, thermal and chemical stimuli.[36] They are thought to possess at least one neurotransmitter, **Substance P**, for influencing the cells they connect with in the spinal cord. Application of **capsaicin** to the skin produces pain mediated specifically through the C nociceptors. A special receptor system in their sensory nerve endings is activated which causes intense electrical discharges in the axons of the C neurons. This sensory input to the spinal cord subsequently elicits a variety of reflexes depending on the site and intensity of the original capsaicin stimulus. Continued application of the noxious stimulus leads to desensitisation of the C nociceptors by a variety of processes, which include depletion of Substance P from the affected sensory neurons. This knowledge has allowed capsaicin to be used in a number of medical applications such as the relief of chronic pain, especially since the receptors which respond to capsaicin have been found in other sites in the body.[37] Nevertheless, it is not too difficult to imagine how such knowledge might be misused.

Mind control

It will be recalled from the SIPRI studies discussed earlier in the chapter that the US Army had an interest in a wide range of incapacitating chemicals. In April 1960 the US professional journal *Army* carried an article entitled 'Mickey Finn on the battlefield'. It was written by the commander of the Army's Chemical Warfare Laboratories who explained that:[38]

> Disabling chemical compounds fall into two broad classes: those which temporarily immobilise the mental faculties – in effect, neutralising a person's will to resist; and those which produce physical distress for a significant period, such as severe discomfort, anesthesia, paralysis or immobility.

Another article of the same period, also written by members of the Chemical Warfare Laboratories, was headed '"Off the rocker" and "On the floor": Non-lethal chemical agents'. It attempted to differentiate between agents which broadly affected thinking and those which prevented action by a person.[39] Whilst such differentiation is far from clear-cut, it is apparent that there was an interest in finding means of attacking the functions of the human brain. Indeed, it appeared from the commander's article that the Chemical Corps had been studying new drugs since the mid-1950s.

As the Director of Research at the Chemical Warfare Laboratories had explained in 1956, this involved searching for new chemicals which might have desirable properties through literature surveys, screening newly-discovered natural products, and seeking help from industry and universities.[40] Necessarily, it would also have involved attempts to synthesise modified forms of substances found to have interesting properties, and tests of such modified substances on both animal and human subjects.

We can obtain an insight into this programme of research through recent disclosures of the extent of human experimentation during the Cold War period. General Accounting Office (GAO) testimony to Congress in 1994 revealed that:[41]

> From 1952 to 1975, the Army conducted a classified medical research program to develop incapacitating agents. The programme involved testing nerve agents, nerve agent antidotes, psychochemicals, and irritants . . . .Army documents identify a total of 7,120 Army and Air Force personnel who participated in these tests, about half of whom were exposed to chemicals . . .

Additionally:

> During the same period, the Army Chemical Corps contracted with various universities, state hospitals, and medical foundations to research the disruptive influences that psychochemical agents could have on combat troops . . .

The report also stated that the Air Force had experiments with LSD (lysergic acid diethylamide) carried out on about 100 people at five universities. More significantly, however, the CIA was also involved:

> . . . from 1953 to about 1964, the CIA conducted a series of experiments called MKULTRA to test vulnerabilities to behavior modification drugs. As a part of these experiments, LSD and other psychochemical drugs were administered to an undetermined number of people without their knowledge or consent . . .

This was part of what John Marks, in his authoritative study,[42] called the CIA's search for the 'Manchurian Candidate'. According to Marks, the point was that for two decades from the early 1950s:

> . . . the Army would invest huge sums in schemes to incapacitate whole armies with powerful drugs. But the CIA clearly pulled far into the lead in mind control. In those areas in which military research continued, the Agency stayed way ahead . . .

Marks argued that the CIA was always at the cutting edge of research and sponsored the majority of the most questionable experiments. This background and potential input to the Chemical Corps activities need to be borne in mind.

The progress of the US Army's research can be followed through the annual reports of the Chemical Corps. Noting that in 1949 a preliminary report, entitled *Psychochemical warfare – A new concept of war*, had been produced, the 1955 report pointed out that there had been increasing interest in those agents which:[43]

> . . . affect adversely the human mind, and/or produce physiological effects which in turn cause psychological disturbances . . .

Three classes of compound had been chosen for investigation:

> . . . lysergic acid and its indole derivatives, the tetrahydrocannabinol series, and the phenethylamines and related substances.

A new sub-project, '4-08-03-016-05, Psychochemical Agents', had been established to discover and develop agents. The military characteristics desired were such that the agent should:

> . . . act promptly, preferably in less than one hour; that it was desirable, but not essential, that it have no permanent effect; that it should have a potency at least equal to nerve gases; that it should have a low toxicity, be stable in storage . . .

Significantly, in relation to conceptions of the agent's potential use, it should also be 'capable of being disseminated from airplanes under all environmental conditions'. The objectives sought, in the effects of the agents, present a staggering list ranging from delusion to suicidal tendencies and through nausea to urticaria.

These points from the Chemical Corps annual report are hardly surprising when we consider the conclusions of a special *ad hoc* study group on 'Psychochemical Agents' set up by the Assistant Secretary of Defense (R&D), chaired by Harold Wolff, and completed between June and December 1955. This study group's report clearly recommended actions of the kind beginning to be undertaken in earnest by the Chemical Corps. Significantly, of the six members of the group, one was a retired military officer, and three others appeared to have been highly active in work of this nature for the CIA – rather confirming Marks' view of the leading role of the agency in the area of psychochemical research.[42] One of the appendices to the study group's report provides a particularly clear idea of the basis on which work for project 4-08-03-016-05 was undertaken and it is reproduced here as Appendix

II. The contents of this appendix will be considered later but the spirit of the times is, perhaps, well conveyed in its concluding paragraph:[44]

> . . . Psychiatrists, pharmacologists and biochemists are actively investigating the problem of the mechanism of mental illness, and as these studies progress leads to more effective psychological agents will undoubtedly become available. As this occurs, our program will be modified accordingly. The pharmacological and clinical studies . . . certainly point in the direction of a better understanding of these matters, *and we therefore look forward with a good deal of optimism regarding the possibility of fulfilling military requirements in this field*. [author's emphasis]

In that context it is perhaps possible to understand how experiments were carried out by the CIA on people without their express permission.[42]

The Chemical Corps annual report for 1956 records that the agenda for the Eleventh CBR Tripartite Conference with Canada and the UK included an item on 'Operational aspects of the employment of psychochemical agents'. It also records that, as a result of the recommendations of the study group on psychochemical agents, Van M. Sim was given responsibility for the CW Laboratories' clinical research programme and the authority to use human volunteers was received in the middle of the year. Twenty-two chemicals were tested for their effects on animals.[45]

By 1957 it had been decided that the agents under study would no longer be called psychochemicals but 'K-agents' instead. Interest continued to focus on the same three groups of agents.[46] The New York State Psychiatric Institute had investigated the clinical effects of mescaline and its derivatives. The institute had tested six derivatives and the Chemical Warfare Laboratories had tested 35. The results demonstrated that 'the doses needed to achieve mental effects were too large'. Previously, Tulane University and the New York State Psychiatric Institute had held contracts to investigate LSD. The University of Maryland Psychiatric Institute was awarded a contract in 1956 to form a research team to investigate LSD. The Second Army provided volunteers and tests were carried out at the Army Chemical Center. As previously suggested by the study group on psychochemical agents, the aim of this research on LSD was to:

> . . . determine if it is practical in impairing the ability of officers to command troops, the ability and willingness of soldiers to obey officers, and the ability of groups of men to perform complicated military tasks such as operating guided missile emplacements and operating tanks.

The University of Michigan was investigating a third K-agent. This was a derivative of tetrahydrocannabinol, with a code symbol of EA 1476. Tests were being carried out on cats and dogs, but there were concerns that the substance could bring on pathological changes.

The report for Fiscal Year 1958 stated that the Corps had been studying K-agents for seven years. Since authorisation had been received for such experiments in 1956, several experiments had been carried out to ascertain the effects of a derivative of lysergic acid on human subjects. Again as suggested by the study group on psychochemical agents, a test had been carried out on a squad of men undergoing routine training. The report concluded, from the behaviour of the experimental subjects that:[47]

> It seems possible that the K-agents, if brought to the point of military usefulness, could be valuable in softening enemy troops, causing them to lose their morale and judgement, putting them in an euphoristic state where they have no interest in warfare, making them halt any offensive action, and prevent them from resisting attack and capture . . .

Naturally, in such a state troops would also be unable to defend themselves effectively if attacked with lethal weapons.

The Chemical Corps report for Fiscal Year 1959 noted that during the Thirteenth Tripartite Conference in September 1958 there was agreement on several major points. One of these was that, 'all three countries should concentrate on the search for incapacitating and new type lethal agents'.[30] By the time of the report for Fiscal Years 1961–62, it appeared that success had been achieved. Under the heading 'Incapacitating Agent BZ and Munitions', the report declared that:[48]

> An incapacitating agent designated BZ was chosen and standardised during the period under consideration . . .

We can now finally turn to this weaponised agent as an illustration of what we might see in the future.

Agent BZ

In the early 1960s the US Army was receiving a large number of substances for testing each month through its contacts in industry. Hoffmann-La Roche sent the first sample of 3-quinuclidinyl benzilate, or BZ, to the army. Dr Van M. Sim had decided to test chemicals on himself before trying them on others. He is reported to have had little trouble with LSD, but stated that BZ had a very marked effect:[49]

> . . . It zonked me for three days. I kept falling down and the people at
> the lab assigned someone to follow me around with a mattress . . .

It has been suggested that the Army did much more testing of BZ on human subjects than of LSD. An estimated 2,800 soldiers are said to have been exposed to BZ between 1959 and 1975. The Army liked the agent because it was cheap to produce, could be spread as an aerosol and would both knock someone 'off his rocker' and put him 'on the floor'.

We can get an idea of the military rationale behind the Army's wish to have BZ in addition to CS from an analysis of potential roles and missions for incapacitants carried out in 1964. This argued that the requirements were for:[50]

> . . . an agent with immediate onset and up to 30 mins duration, and an
> agent with less than 30 mins onset and up to several days' duration . . .

Since CS has a duration of some ten minutes and BZ an onset of about two hours they were considered a less than ideal pair, but the analysts believed:

> . . . that the capabilities of CS and BZ can be exploited in many
> situations and that CS and BZ weapon systems have great potential
> utility . . .

This statement was made before the massive use of CS in Vietnam. It should be noted that, looking to the future, the analysis also covered a putative agent similar to BZ in all respects other than requiring just one quarter of the dose.

We have detailed information on the militarily significant characteristics of BZ from the *Joint CB Technical Data Source Book* of 1972.[51] Agent BZ was described in the sourcebook as the standard incapacitating agent of US armed forces, but again it was noted that other improved chemical agents were under development and that one or more of these might replace BZ at a later date. BZ was described as a psychophysical and anticholinergic incapacitant. It was disseminated in an aerosol from a pyrotechnic munition with the primary route of entry to the human body being through inhalation.

One problem with BZ seems to have been the considerable variation in estimates of the amount necessary to affect human beings. An official value of 112 mg min/m^3 to incapacitate half of an exposed population was given. Estimates of the dosage required to kill one per cent of an exposed population ranged from 3,800 to 40,000 mg min/m^3. There was thus perceived to be a large safety margin between the effective and the lethal dose. The effects of BZ were dramatic. Four examples were given of the possible degrees of incapacitation depending on the dose of BZ received. Following a mild dose people would experience slight blurring of vision, sleepiness and some mental retardation. With a moderate dose such symptoms would increase and be accompanied by fleeting illusions and hallucinations. A severe dose would disorganise a person so completely that military operations were impossible:

> . . . Hallucinations, confusion, hyperactive disorganised behavior, incoherent speech, and disturbances in mentality and attention characteristically appear following an early period of deep sleep or stupor. Peak heart rates reach 95 to 110 . . .

A maximal dose would push peak heart rates to between 110 and 140 beats per minute and:

> . . . rapid onset of stupor occurs, often preceded by a period of agitation. Stupor is followed by a protracted period of sleeplessness, disorganised random behaviour, continual hallucinations, and sometimes impulsive outbursts of fear and anger based on misinterpretation of surroundings . . .

Effects appear to peak between 3–8 hours after the onset of symptoms and, if untreated, to last three to four days until recovery is complete. Figure 5.3 reproduces an illustration from the sourcebook showing the chronological sequence of effects (in hours) caused by BZ. Physostigmine salicylate (eserine) was found to reverse the effects of BZ, but this is a powerful drug in its own right which could itself cause problems if administered in excess of what was required.

It should be understood that these results were obtained by experimentation on human subjects. As the sourcebook indicated:

> Time effects following BZ administration have been obtained from investigations conducted by the University of Pennsylvania on subject populations at Holmesburg Prison and from the Clinical Research Department at Edgewood Arsenal . . .

Figure 5.3 Chronological Sequence of BZ Effects*

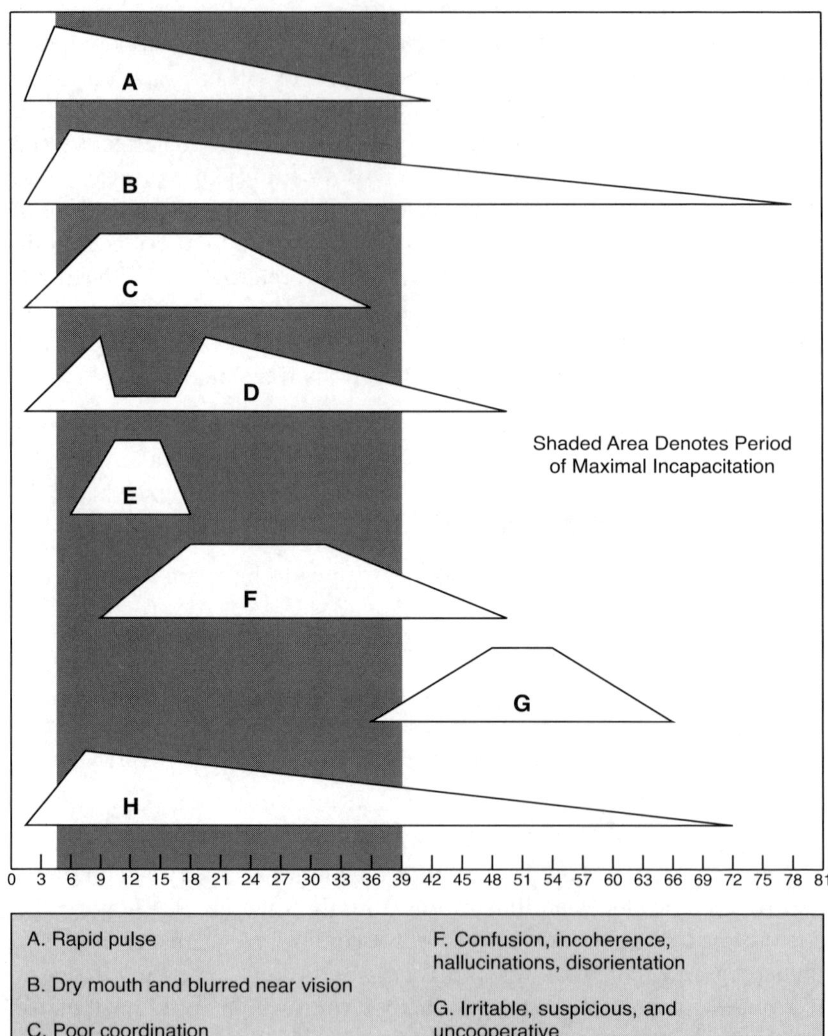

* From reference 51

This work also had a real military output. As can be seen from Table 5.5, a range of munitions had been developed for the dissemination of BZ. With an onset time for the effects of BZ of some two hours after inhalation, the munitions appear to be geared more to the longer-range, larger-area coverage end of the spectrum of uses that CS had in Vietnam. For example, an M43 cluster contained 57 M138 bomblets. Each bomblet contained four M7 canisters, each of which held 170 grams of the pyrotechnic mixture. Thus there were in total 228 canisters carrying 39 kilograms (86 pounds) of agent fill. With an estimated dissemination efficiency of 60–70 per cent, the required incapacitating dose would have been achieved over an area of 8,800 square metres.

Table 5.5 BZ munitions*

Munition	Weight of agent fill (g.)	Submunitions
M6 canister	140	
M7 canister	170	
M16 generator	5,900	42 each M6 canisters
M138 bomblet	680	4 each M7 canisters
M43 bomb, cluster	38,800	57 each M138 bomblets
M44 generator, cluster	17,900	3 each M16 generators
CBU-5/B cluster	38,800	57 each M138 bomblets (an M43 modified for high performance aircraft)
E157 cluster	14,300	21 each M138 bomblets
BLU-50/B bomblet	22	
CDU-9/B canister, cluster	700	32 each BLU–50/B bomblets
CBU-16A/A bomblet	28,200	40 each CDU–9/B canister clusters

* From reference 51

What is obvious is that with BZ we have moved far beyond a simple harassing agent. We are dealing with a substance that, in order to produce effects such as hallucination, must be affecting the central nervous system. We know that BZ is labelled as an anticholinergic and we know that there are some cholinergic systems in the brain. Moreover, it is clear that BZ is only one of several hundred such anticholinergic compounds which were investigated in the United States at the time.[52] Today, military scientists studying BZ, in order to dispose safely of the agent, caution that it is a harmful material requiring 'extreme care' to be taken in laboratory work.[53] Nevertheless, there are numerous papers in the current literature concerning the effects of BZ (now termed QNB) on the nervous system, since it has been found to be a very specific agent for binding the muscarinic class of cholinergic receptors. It is perhaps not surprising to discover that even during the 1980s some of the work on the effects of BZ on the nervous system was still being funded by the US Department of Defense.[54] There have recently been reports that BZ was used by Serbian forces in Bosnia during the later stages of the civil war there. According to one account, shells filled with BZ were used to break the defence of Zepa. The shells apparently fell with a dull thud and not with an explosion. Then, one victim said:[55]

> . . . suddenly we could not breathe. We stumbled out of the trenches. . . . Men fell down and could not get up. Our noses, throat and lungs felt like acid . . .

The soldier continued:

> Those who were badly hit by the gas spent the next three days lying flat on their backs. They were in constant pain, nauseous and had trouble moving . . .

Another soldier reported that Serbian troops wearing gas masks stormed the trenches following the gas attacks. He said that it was impossible to hold on to the front line of trenches:

> . . . Soldiers who were badly gassed needed five days to recover. . . . If they got back to the front lines too early . . . they were next to useless. No-one died from the gas, but whatever this stuff was, it was powerful and effective.

Though it is never wise to jump to conclusions over uncorroborated reports from zones of conflict, some experts have suggested that these effects might well have resulted from the use of an anticholinergic such as BZ.

What is required now is a fuller understanding of the structure and function of the human nervous system, the subject of the next chapter. We then need to know something of the status of modern work on the chemistry of the brain in order to appreciate the importance of current work on incapacitants. These two topics are the subjects of Chapters 7 and 8 respectively.

.6.

THE HUMAN NERVOUS SYSTEM

The last chapter gave a brief description of the structure of a neuron or nerve cell and of how it transmits information across the synapse to another neuron. It also described how input from a sensory neuron could affect the output from a motor neuron in the spinal cord and produce a reflex action. This chapter emphasises the scale of change ongoing in our understanding of how the whole human nervous system functions. As Richard Thompson noted in the introduction to his review of recent research,[1] 'The research and thought that led to our current understanding of neuron and brain is one of the supreme achievements of science in the 20th century'. He also emphasised the significance of this research:

> . . . The United States Congress has declared the 1990s the 'Decade of the Brain' with good reason. We are poised on the threshold of a truly deep understanding of the human brain and mind.

Traditionally, the structure of the nervous system was thought of only in anatomical terms. The function of the system was then considered in relation to the anatomical structures. But now, as Roger Barker noted in his overview of neuroscience:[2]

> . . . it is perhaps more meaningful to consider the central nervous system (CNS) *neuropharmacologically* [in terms of its neurons and transmitters] rather than in the classical anatomical and functional terms . . . [author's emphasis and addition]

Parkinson's disease, to cite an example, is now best understood in terms of the disruption of certain chemical circuits in the brain. However,

95

those seeking more effective non-lethal chemical agents could also use this new research to advantage.

Basic Anatomy of the Nervous System

The human nervous system consists basically of the **central nervous system** or CNS (brain and spinal cord) and the **peripheral nervous system** (Figure 6.1). Input from peripheral sensory receptors is processed within the CNS after it is received through afferent (travelling towards the CNS) pathways. This is illustrated on the left-hand side of the Figure. Output to muscles is then sent from the central parts of the system through efferent (away from the CNS) pathways. A further division of labour can be seen in the Figure. The human nervous system is concerned not only with the control of skeletal musculature (arms, legs etc.), but also with visceral musculature (in internal organs such as gut, heart etc.) and glands. Skeletal muscles are controlled by the **somatic nervous system**, and internal organs by the **autonomic nervous system**.[3] 'Special' sensory receptors in Figure 6.1 refers to complex receptor systems such as eyes and ears.

The autonomic nervous system

An insight into the chemical organisation of the nervous system may be gained by looking at the autonomic nervous system (ANS) in more detail. As its name implies, this part of the human nervous system is concerned with the complex regulation of bodily functions essential for survival but not normally under conscious control (ie involuntary) such as heartbeat, digestion and excretion. There is a basic difference between the somatic nervous system which controls our voluntary motor systems (skeletal musculature) and the ANS which controls involuntary systems; in the former a motor neuron at the centre can directly innervate or supply a muscle, whereas in the ANS the message from a central neuron is directed first to a peripheral autonomic ganglion (collection of nerve cells) located either close to the spinal cord or close to, or within, the target organ. Nerve cells in such peripheral ANS ganglia then innervate the effector organs.

The autonomous nervous system itself has two divisions: **sympathetic** and **parasympathetic**. These two systems usually act in opposite ways: if one is excitatory in its effects on an organ, the other is likely to be inhibitory. One way to think about this is by remembering that the sympathetic division is active in producing the 'fight or flight' response

Figure 6.1   An Overview of the Function of the Nervous System*

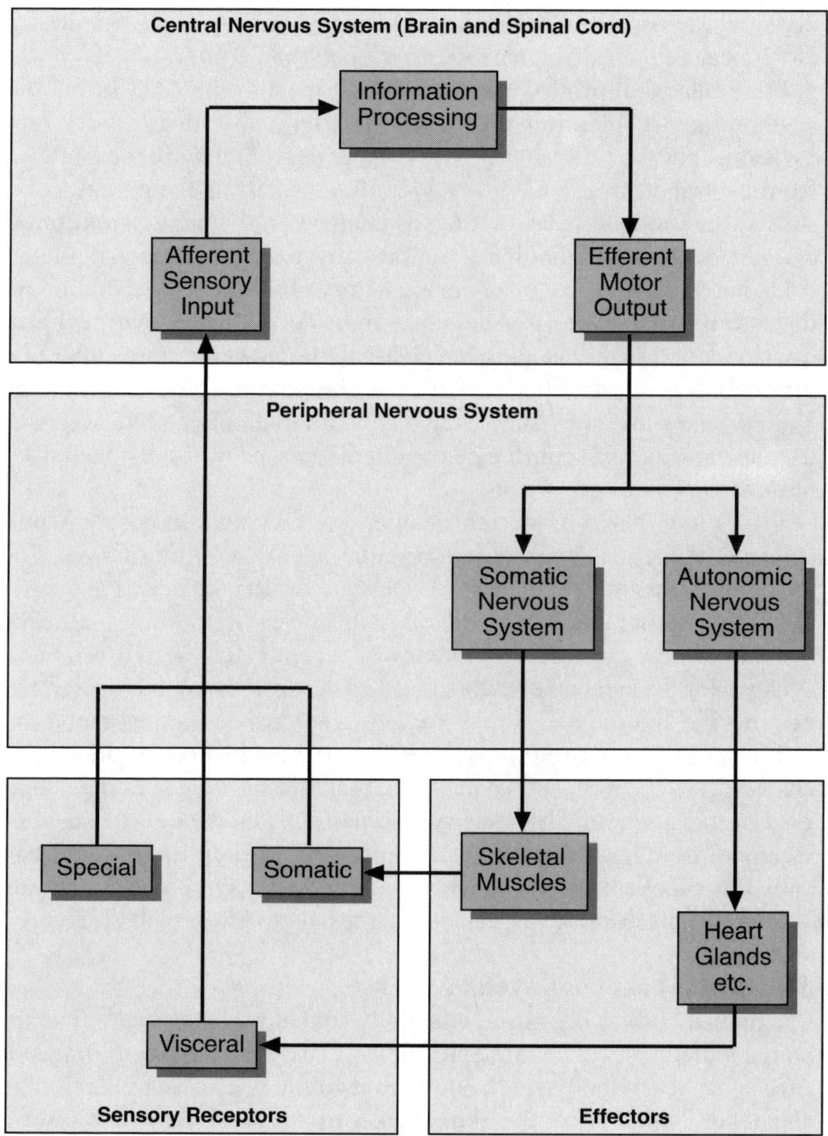

* After reference 3

to a sudden potential threat: heightened awareness and metabolism, and reduced vegetative functions such as digestion. The parasympathetic division, in contrast, is active when we are relatively quiescent after a large meal and our digestive system is operating strongly.

From our point of view the interesting aspect of the ANS lies in the systematic organisation of the transmitter substances in its two divisions. All the pre-ganglionic axons (or nerve fibres) in the ANS (ie from central neurons) have **acetylcholine** (ACh) as their transmitter substance, and its effect is always excitatory. In the **parasympathetic** division the post-ganglionic nerve fibres are also cholinergic and release ACh, but its effects may be either excitatory or inhibitory, depending on the nature of the receptors of the target cells. In the **sympathetic** division, on the other hand, most post-ganglionic nerve fibres release a different neurotransmitter called **norepinephrine** (NE). Outside the United States this substance is also called noradrenaline, but we shall use the name norepinephrine here. Norepinephrine is usually excitatory in its effects on target organs.

It was mentioned in the last chapter that receptors in post-synaptic cells at cholinergic synapses can be either nicotinic or muscarinic. For example, parasympathetic cholinergic synapses on the heart have muscarinic receptors and the effect of ACh release is to slow the heart. Similarly, there are different classes of receptor for the NE-activated synapses of the sympathetic division. The synapses using NE are termed **adrenergic** and the major receptor sub-types are designated alpha (α) and beta (β). These receptor types will be treated in more detail in later chapters. For the moment we may note that this knowledge of the chemistry of the autonomic nervous system has allowed the development of means to treat numerous medical conditions. Examples are the drugs called 'β-blockers' which, since the action of the sympathetic nervous system stimulates the heart, can be used to help to lower blood pressure.[4]

The central nervous system

The human body is organised bilaterally so that we have predominantly paired structures – cerebral hemispheres, eyes, ears, hands etc. – one on each side of the body. Its basic organisation is also segmental. The spinal cord which, with the brain, makes up the central nervous system (CNS), and whose structure was briefly discussed in Chapter 5, is divided into a series of segments centred on the vertebrae, each half segment having a dorsal and a ventral nerve root as shown in Figure 5.2. This kind of segmental organisation is typical of nervous systems in other segmented animals such as worms and insects, but human nervous systems are characterised by the extremely large number of individual nerve cells.[1]

Our aim here is to grasp the important features of the structure and functions of the human brain for, as will become apparent, those interested in creating new calmative agents are mainly concerned with disrupting its function. The human brain is similar in structure to that of other primates, particularly the great apes, to which we are closely related.[5,6] Most people are probably familiar with a lateral or side view of the brain, dominated by one of the bilateral hemispheres of the **cerebrum** or **cerebral cortex** (Figure 6.2). Most have also probably

Figure 6.2   Simplified Lateral View of the Brain*

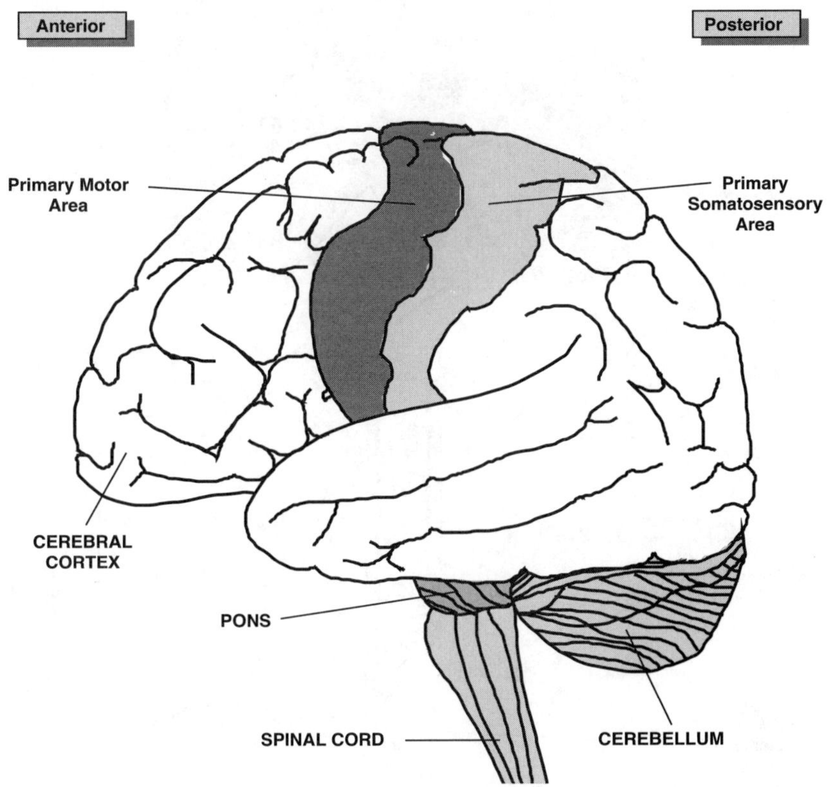

* After reference 3

heard of the primary motor and sensory areas lying on either side of the central sulcus or groove in the heavily folded cerebral cortex. If we consider a schematic plan view of the brain and spinal cord, concentrating on the connections to and from these primary areas, we can get an initial overview of how the whole central nervous system is organised (Figure 6.3).

Figure 6.3   Simplified View of Sensory and Direct Motor Pathways*

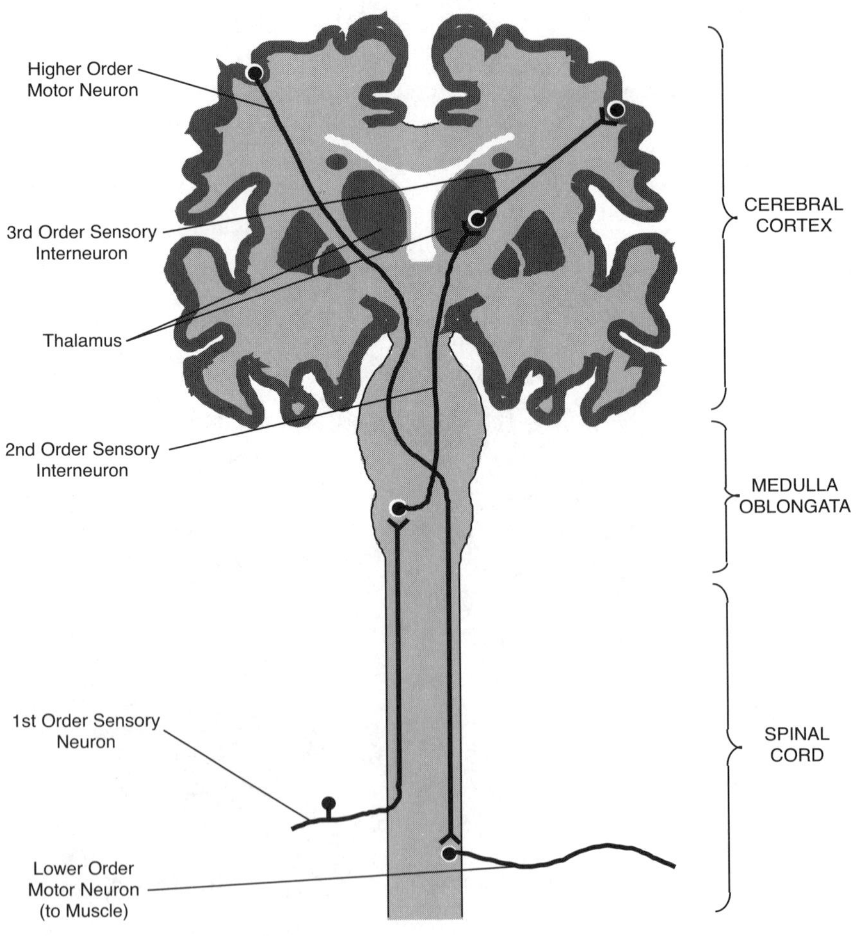

* After reference 7

As noted previously in regard to the autonomic nervous system and the sensory cells in the dorsal root nerve of the spinal cord, collections of neurons outside the CNS are called **ganglia**. Within the CNS, collections of neurons with a common function are called **nuclei**. Again, outside the CNS, collections of nerve fibres travelling as a group are simply called **nerves**, for example the cranial nerves which innervate the face. Inside the CNS, collections of nerve fibres travelling as a group are called **tracts** (or **commissures** if they link the two halves of the brain). So we should think of the CNS as an information-processing system composed of nuclei (of nerve cells) linked by tracts (of nerve fibres).

The route followed by a nerve impulse through the nervous system is called a nerve pathway.[7] We can follow the hypothetical pathways set out in Figure 6.3, starting with the input from a first order sensory neuron at the bottom left-hand side of the figure. This neuron is shown synapsing on to a second order interneuron in the **medulla oblongata**, at the base of the brain. It will be noticed that in this region of the brain the fibre then crosses over to the other side. It terminates in nuclei of a structure called the **thalamus**. This is the principal relay station for sensory input from the spinal cord to the cerebral cortex. Third-order interneurons then convey the incoming information to the primary sensory cortex. On the other side of the figure a higher order motor neuron in the primary motor cortex is shown sending an axon down towards the spinal cord. This again crosses over in the basal part of the brain and synapses on to a motor neuron in the spinal cord which directly innervates a muscle. In addition to this *direct* pathway, lower order motor neurons in the spinal cord may be affected by much more indirect influences involving other nerve cells in, for example, the **cerebellum**. This is the second largest structure of the brain and is located in its posterior lower region.

The simplest way to understand the relationships between the major structures of the human brain is by reference to their development in the embryo. The central nervous system starts out as surface tissue which becomes infolded to form a hollow longitudinal tube filled with fluid. In the first three weeks of growth the head end of this neural tube expands, through enlargement of the fluid space, to create three primary brain vesicles: the prosencephalon (**forebrain**); the mesencephalon (**midbrain**); and the rhombencephalon (**hindbrain**). After six weeks, the forebrain and the hindbrain have each divided further into two secondary brain vesicles. The forebrain has given rise to the telencephalon and the diencephalon, while the hindbrain has divided into the metencephalon and the myelencephalon (Table 6.1). Then, eventually, the telencephalon of the forebrain gives rise to the cerebrum with its cerebral hemispheres which cover the top and side surfaces of the brain. The midbrain produces the tectum and the tegmentum. The metencephalon – that part of the original hindbrain nearest to the midbrain – gives rise to the large cerebellum and the **pons** (or bridge) which links it to the rest of the system. The myelencephalon of the hindbrain, which is the closest of all these structures to the spinal cord, forms the medulla oblongata.

Table 6.1   Development of the human brain

Primary brain vesicles	Secondary brain vesicles	Major derived brain regions*
Prosencephalon (forebrain)	1. Telencephalon	1. Cerebral hemispheres 2. Basal ganglia (striatum)
	2. Diencephalon	1. Thalamus 2. Hypothalamus
Mesencephalon (midbrain)	Mesencephalon	1. Tectum 2. Tegmentum (substantia nigra)
Rhombencephalon (hindbrain)	1. Metencephalon	1. Cerebellum 2. Pons (locus coeruleus)
	2. Myelencephalon	Medulla oblongata

* Some structures of these regions, referred to in the text, are given in brackets

As can be seen in Figure 6.2, a cerebral hemisphere and the cerebellum dominate a lateral view of the brain. These structures are covered with a layer, or cortex, of grey matter (nerve cells). The nerve cells of the cerebral hemispheres are concerned with complex functions such as language, understanding and memory as well as originating complex motor outputs. The cerebellum is involved in the fine adjustment of motor activities. If the cerebral hemispheres and the cerebellum are stripped away from the rest of the brain, the remaining smaller structures are exposed (Figure 6.4). The original diencephalon of the forebrain gives rise to the thalamus, which is the final relay centre for sensory information, as we have already noted. Below it lies the **hypothalamus** which contains neurons concerned with autonomic functions, the emotions and hormone production. It is linked to the **pituitary gland** which plays a key role in the hormonal control of bodily functions.

Figure 6.4   Brain Structures*

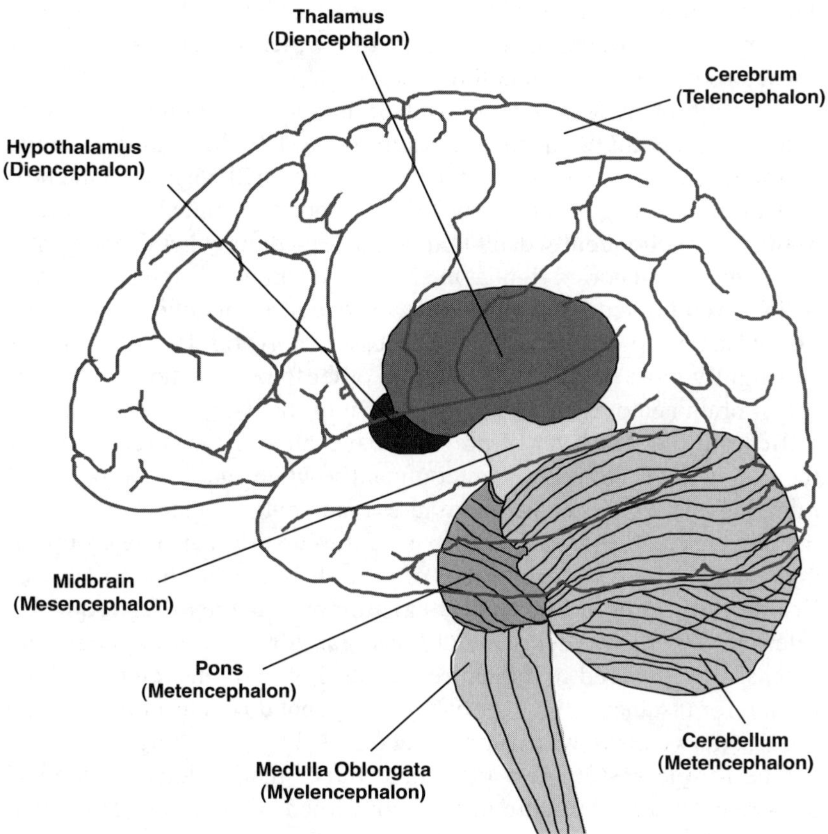

* After reference 3

The midbrain is involved, among other things, in the processing of visual and auditory inputs and in the generation of involuntary movements such as the 'startle' response to a bright light or noise. In the hindbrain the pons relays information to the cerebellum and other parts of the brain. The pons also contains neurons involved in the regulation of breathing. The medulla oblongata of the hindbrain again has relay functions and also contains the motor neuron centres that control autonomic functions such as heart beat and the basic pace of breathing. A distinction can be made between the embryonic telencephalon of the forebrain and the diencephalon, mesencephalon and parts of the rhombencephalon which together eventually constitute the **brain stem**.[8] In general terms, the brain stem is concerned with more basic functions while the higher centres of the brain (cerebral hemispheres and cerebellum) control more complex activities which include those under conscious monitoring and control.

A wealth of more detailed information is available on the structure and function of the elements of the nervous system, for example: on the various regions of the cerebral cortex concerned with vision; on the processing of language which, unusually in our fundamentally bilateral brain, is most frequently dealt with in the left cerebral hemisphere; and on the prefrontal cortex right at the front of the brain, which is known to be involved in predicting the future consequences of actions or events. For present purposes, though, it is necessary only to select a few points for slightly more detailed treatment in order to have some understanding of brain anatomy to set alongside that of the intricate chemistry.

Beneath the thin layer of grey cellular matter of the cerebral cortex there is a large volume of predominantly white matter composed of radiating nerve tracts and commissures containing vast numbers of connecting nerve fibres. Within this part of the cerebrum also lie groups of nerve cells. These various groups of cells should, according to our earlier definition, be termed **cerebral nuclei**, but they are frequently referred to by their older name of **basal ganglia**. The basal ganglia are extensively involved in the indirect control of body movements. They do not, for instance, initiate or halt walking, but do control the rhythmic movement we make with our arms and legs during walking.

The **limbic system** is a functional rather than strictly anatomical grouping of brain structures. It is concerned with setting emotional states and generating drives, linking conscious cerebral activities with autonomic functions and memory. The system includes cerebral structures such as the **hippocampus** – so-called by early anatomists because its shape resembles that of a sea-horse – and some parts of the diencephalon, such as the hypothalamus, located near the cerebrum. The limbic system is also linked to the **reticular formation**, a dense network of nerve cells which runs throughout the brain stem. A final point to note is that the midbrain contains a system called the **substantia nigra**. This area of nerve cells is darkly pigmented.

Chemical Circuits in the Brain

According to Richard Thompson's review:[1]

> . . . The study of chemical substances and circuits in the brain and drug actions on these circuits and neurons began only a few years ago but has greatly expanded to become *the largest field of neuroscience.* [author's emphasis]

We can get an idea of the increasing possibilities for control of the nervous system generated by this advancing research by looking at two well-known diseases.

Parkinson's disease was first described by an English doctor in 1817. Today its main symptoms are recognised to be a distinct tremor when the person is at rest, rigidity of the limbs, and bradykinesia – slowness of movement and inability to initiate movements.[9] As the disease progresses, normal postural reflexes also begin to fail and people with the disability are at risk of being unable to correct a minor stumble before falling. The prevalence of Parkinson's disease in the population increases with age. In the United States it is about 30–50 per 100,000 in the fifth decade of life and between 300 and 700 in the seventh. A clue to the origins of the disease arose from the fate of a group of drug addicts who had unfortunately injected themselves with a 'designer' drug called MPTP.[10] This irreversibly produced the standard characteristics of Parkinson's disease, because the drug preferentially killed nerve cells which utilised **dopamine** (DOPA) as a transmitter substance. It had actually been noted during the 1950s that, in people with Parkinson's disease, parts of the brains were profoundly depleted of the normal level of dopamine. This discovery had led to efforts to restore the lost dopamine in the brain. By the late 1960s it had been discovered that, if given in large quantities, a precursor of dopamine called **levodopa** (L-DOPA), would cross over from the blood to the brain in sufficient quantities to reverse the effects of the disease.

Three major types of dopamine circuit have been discovered in the brain.[1] All have dopamine-containing cells located in the brain stem which send their axons to other parts of the brain. One type of circuit involves neuron cell bodies and axons located in the hypothalamus. Some of these cells act directly as part of the hormonal control system.[9,11] A second type of circuit involves cell bodies in the midbrain that send their axons into the frontal area of the cerebral cortex and other parts of the limbic system. These cells are thought to be involved in major brain dysfunctions.[11] The third type of dopamine circuit is the best known. Many of the darkly-pigmented nerve cells of the substantia nigra have dopamine as their transmitter substance. These cells contain approximately three-quarters of all the dopamine in the brain and their axons travel directly to the basal ganglia which, it will be recalled, are involved in the detailed control of movement. There are about 400,000 dopamine neurons of the substantia nigra which nearly all send their axons to a structure termed the **striatum**. This structure contains important basal ganglia cell groups. The predominant effect of the dopaminergic input is to inhibit the cells in the striatum. It has been shown that in Parkinson's disease:[8]

> . . . The cardinal *pathological* feature is the loss of neurons from the substantia nigra. At least 80 per cent of . . . neurons have already been lost before symptoms appear . . .

In the early stages of the disease loss of these cells is compensated for because the remaining cells produce more dopamine and there is an increase in the number of receptors for DOPA on the target cells. However, the system is eventually overwhelmed by the disease. Loss of the inhibition on the target cells produces excessive excitation of motor neurons. Initiation of movement becomes difficult because opposing muscles do not relax in turn as in an unaffected individual. The inhibition on the target cells can, however, be restored by taking the drug L-DOPA.

Unfortunately, about half the people successfully treated with L-DOPA experience severe complications after about five years. These consist either of sudden intermittent and transient loss of the anti-Parkinson effect of the drug or the development of involuntary movements of the trunk or extremities. Even here, however, all hope is not lost. Complex models of the neuronal control circuits involved have been built up[2,8] and, as a remarkable 1995 television documentary demonstrated, careful destruction of one of the minute basal ganglia involved stopped the terrible writhing movements of one particular patient.[12]

Alzheimer's disease is the commonest cause of dementia. There are about two million sufferers from the disease in the United States and it causes many deaths.[3] Its early symptoms usually appear in people between 50 and 60 years of age and there is large-scale damage to the cerebral cortex in affected people. One of the striking features of Alzheimer's disease is loss of memory, and it is known that the hippocampus, important in short-term memory, is among the structures damaged.[8] Moreover, there is evidence that cholinergic input to the hippocampus is important in learning and one of the characteristic features of Alzheimer's disease is loss of cholinergic input to the hippocampus and the cerebral cortex. The cholinergic cell bodies lie in the basal part of the forebrain within the basal nucleus of Meynert[8] and in the disease, 'Up to 90 per cent of the ACh neurons are lost from the basal nucleus, as are their projections to the cerebral . . . cortex'. The mechanisms involved in these changes are not yet well understood. It may be that the initial damage is to the target cells the cholinergic neurons innervate and then, with the loss of that connection to the target cells, the ACh neurons themselves degenerate.

There is a clear connection between such studies of the mechanism of Alzheimer's disease and the very rapid advances taking place in molecular biology, a point that we shall be taking up in the next chapter. With Alzheimer's disease, it has been known for some time that certain families have a particular susceptibility. The journal *Science* reported in 1995 that:[13]

> . . . less than two months ago, researchers reported the discovery of a gene on chromosome 14 that is responsible for 80 per cent of the cases of familial Alzheimer's disease, a particularly virulent form of the disorder that can strike as early as age 40.

The report went on to note:

> . . . the discovery of a gene on chromosome 1 that may account for most of the remaining familial Alzheimer's cases. Even more remarkable, the new gene is closely related to the chromosome 14 gene . . .

From this basis, an unravelling of the mechanism by which the degeneration takes place, and the discovery of the role of the damaged cholinergic neurons, might happen quite quickly.

An indication of the possibilities inherent in the coming together of modern molecular biology and neuroscience comes from a recent investigation of the role of acetylcholine in the cerebral cortex of rats.[14] The experimenters were able selectively to destroy the cholinergic neurons of the basal nuclei involved, and to demonstrate how this severely affected the rats' ability to perform a spatial learning task. They then implanted, in the cortex region deprived of acetylcholine input, genetically modified cells that produced acetylcholine. The rats subsequently showed a significant increase in their ability to perform the same spatial learning. Commentators remarked that, although the precise mechanisms involved needed further clarification, the central message of the work was clear:[15]

> . . . namely that genetically engineered cells can be used to provide a local supply of biologically active molecules to an affected brain region in order to promote functional recovery and repair . . .

Clearly this opens up both new means of studying disease and of developing viable therapies.

Like ACh, norepinephrine (NE), the other transmitter substance we encountered in our brief description of the autonomic system, has well marked circuits in the brain. The NE circuits are unusual because a very small number of brain cells – about 20,000 in total – are involved. Moreover, these cells are located almost entirely in the locus coeruleus of the pons whose name derives from the blue colour these pigmented cells give to the area. Despite the restricted location of their cell bodies, the axons of these NE cells are highly branched and terminate in many parts of the CNS including the cerebral cortex, limbic system, cerebellum and spinal cord. The axons have what are termed 'diffuse' synapses, with terminals located along the length of the distal parts of the axon branches rather than discrete endings just at the tip of each branch. All this suggests a system designed to bathe a target area with transmitter to achieve a generalised effect, and it does appear that the NE system is involved in control of the general state of arousal of the brain.[9]

The distribution of **serotonin** (5-hydroxytryptamine, 5HT) nerve terminals has similarities to that of NE neurons. The cell bodies are located in the so-called **Raphe nuclei** of the midbrain, pons and medulla. In general terms, the serotonin cells of the midbrain innervate the higher brain structures while those of the pons innervate the brain stem and cerebellum and those of the medulla supply the spinal cord. These serotonin-transmitting cells seem to be involved in activities such as sleep and wakefulness, but their exact role is unclear.

It seems almost certain that current research on the human central nervous system will lead to profound changes in the way we view the world. As Paul Churchland has argued:[16]

> . . . Slowly at first, but with compounding momentum, an accessible conception of brain function will spread through both the language and practice of human social commerce . . .

Obviously, the precise knowledge of the functional anatomy and chemistry of the nervous system presaged by the kind of work just discussed could be manipulated for military or political purposes in the future. That is a significant danger we might do well to consider carefully. Of more interest to us now, however, are the capabilities already available to the military, consideration of which requires a small step back in time.

Mechanisms of Incapacitation

As we saw in the previous chapter, the US Army Chemical Corps set out in the mid-1950s to seek effective, non-lethal, chemical agents and eventually ended up with BZ as an interim agent. In the early 1970s SIPRI concluded that at least 12 different mechanisms of incapacitation had been under study.[17] Examples were the creation of hypotension (sudden lowering of blood pressure which produces fainting) through chemicals related to the active constituent of cannabis, and the induction of emesis (retching and vomiting) through harassing agents such as adamsite or staphylococcal enterotoxin B (agent PG) or the injection of apomorphine. It was also known that atropine (the standard antidote to nerve gas) and various bacterial toxins could profoundly disturb the control of body temperature. Various other drugs could interfere with balance, cause paralysis or tremors, or have psychological effects.

The problem with most of these agents was an insufficiently large margin between the effective incapacitating dose and the lethal dose to allow for relatively safe usage. As the SIPRI account emphasised, the search appeared to have narrowed down to seeking effective CNS depressants and CNS stimulants. An example of a depressant was BZ, and of a stimulant, LSD. What is interesting about the discussion of agents in the recent literature is the use of the word 'calmative' almost interchangeably with 'incapacitant'. This would appear to suggest that the principal current interest is in substances which induce sedation or anaesthesia – rendering people insensible.

There has been a long-term interest in 'incapacitation by anaesthetic agents', as evidenced by a contribution to a 1958 Chemical Warfare Laboratories symposium in the United States.[18] The interest lay in general anaesthetic means of rendering an individual unconscious, and not in local anaesthetics. The author pointed out:

> . . . Sedatives, hypnotics, and anesthetic agents are classified pharmacologically as depressants of the central nervous system and in many cases the same drug may produce sedation, hypnosis or anesthesia depending on dosage. The narcotics should be included as a special group in that they will usually produce hypnosis in sufficient dose but are given principally for the relief of pain . . .

While the spinal cord is usually affected, the author stressed that it was the action on the brain which was responsible for anaesthesia. He also noted that:

. . . In general there is a progression of depression from the higher to the lower centers of the brain. . . . The rate and pattern of the progressive depression varies with each of the individual anesthetic agents . . .

The then generally accepted standard four stages of anaesthesia were set out in tabular form (Table 6.2).

Table 6.2 Characteristics of anaesthesia*

Stage	Characteristics		Site of depression
I. Analgesia or altered consciousness	Euphoria; loss of discrimination; reduction of pain; impairment of contact with environment		Depression of sensory cortex
II. Delirium or excitement	Loss of consciousness; occasional uninhibited muscular activity; response to pain may be exaggerated; violent physical movements possible		Inhibition of cortical and sub-cortical control levels
III. Surgical anaesthesia	(a) Depression of centres maintaining respiration and blood pressure	(b) Skeletal muscle relaxation	
	Slight effect through to a marked effect	Slight effect through to a marked effect	Moderate depression of subcortex through to a moderate depression of midbrain
IV. Medullary paralysis	Respiration ceases; low blood pressure; tissue anoxia		Depression of pons and medullary centres

* From reference 18

Clearly, for surgical work the anaesthetised individual has to be maintained carefully at the required state. What is of interest is the author's attempt to set out the requirements for a suitable agent for incapacitating groups of personnel by anaesthesia (Table 6.3). Another Army participant at the symposium had defined the dosage levels required:[19]

. . . Under practical CW [chemical warfare] circumstances . . . these compounds must produce their effect in intravenous doses of 10 mmg./kg. or less. I do not mean that we intend to use these intravenously, but they must have at least that activity. They must produce their effects by inhalation when inhaled for less than a minute in concentrations of 100mmg./l. or less . . .

A review of the chemicals standardly used for anaesthesia (eg nitrous oxide gas), hypnosis (pentobarbital) and narcosis (morphine, methadone) showed that all had drawbacks. However, it was concluded that some drugs 'may be proven to be excellent agents for this purpose' in the future.

Table 6.3   Requirements of a satisfactory agent for incapacitation by anaesthesia*

1. Can be dispersed as gas, vapour or aerosol to be absorbed by inhalation.
2. Adequate potency.
3. Short latent period.
4. Wide margin between initial depression of central nervous system and death.
5. Persistence of action for several hours.
6. Complete reversibility.
7. Preferably non-explosive.

* From reference 18

Since that time the major advances in our knowledge have been in understanding the general distribution of an ever-increasing number of possible transmitter substances in the brain; in an ever more detailed understanding of the mechanism by which such transmitters affect the receptor systems on post-synaptic cell membranes; and in a growing systematic understanding of the action of different drugs on the functions of such transmitters and receptors. It is on the basis of this new knowledge, gained since the early phases of the US Army's incapacitants programme, that today's modern incapacitants are likely to have been developed. This knowledge, then, is central to a proper understanding of what is militarily likely now. In the remainder of this chapter some other potential transmitters will therefore be reviewed; in Chapter 7 transmitter-receptor mechanisms and the burgeoning knowledge of drugs which affect the central nervous system will be discussed.

Other CNS Transmitter Substances

Norepinephrine and dopamine are examples of a group of substances called **catecholamines**. A third member of this group, **epinephrine** (adrenaline), has also been recognized as a probable transmitter. The cells which contain epinephrine are located in the medullary reticular formation and send axons both forward to higher centres and down into the spinal cord. Another amine, **histamine**, which has extensive peripheral distribution and functions (in allergic reactions of the skin and mucous membranes, for example), has also been recognised as a central neurotransmitter. In common with the other amines, the cells of this histamine group have a restricted distribution in the ventral posterior hypothalamus but send long axons forward to innervate the forebrain extensively. These centrally active amines head the list of some major transmitters set out in Table 6.4. We have already dealt with the second major example, acetylcholine. It should be noted, however, that it is exceptionally difficult to demonstrate for certain that a chemical is a functional neurotransmitter, and in most instances this holds true even for ACh. The evidence for many of the other substances is weaker, but most of the major chemicals discussed here are nevertheless widely accepted as neurotransmitters by the scientific community.

Table 6.4   Major central nervous system (CNS) transmitters*

---

1. **Centrally active amines**

        Dopamine (DOPA)
        Norepinephrine (NE)
        5-Hydroxytryptamine (5-HT or serotonin)
        Epinephrine
        Histamine

2. **Acetylcholine**

3. **Centrally active amino acids**

        γ-amino butyric acid (GABA)
        Glycine
        Glutamate
        Aspartate

4. **Centrally active peptides**

        Substance P
        Enkephalins and endorphins

---

* From reference 3

There was inititally resistance to the idea that such simple and ubiquitous substances in the body as amino-acids could be transmitters in the central nervous system. However, there is now good evidence that **GABA** (gamma-aminobutyric acid) is the major inhibitory transmitter for local interneurons in the human brain and that it has inhibitory functions in the spinal cord also. **Glycine** is thought to be the inhibitory transmitter between spinal interneurons and motor neurons, and in much of the reticular formation. There is also good evidence that **glutamate** and **aspartate** function as the principal transmitters for excitatory communication between certain areas of the CNS, particularly through long axons from the cerebral cortex to other deeper areas of the brain.

There are many more substances which may be transmitters in the CNS. Of these we shall consider only the centrally active peptides. **Substance P**, thought to be the transmitter for certain sensory neurons entering the CNS, has already been mentioned. Here we shall concentrate on the fascinating discovery of natural substances in the brain which have similar actions to morphine.

Opioids

The first undisputed reference to the properties of extracts from the opium poppy dates from the 3rd century BC. It was not until the 16th century that the uses, and problems associated with use, of opium were well understood, and not until the last century that the active constituents began to be identified. **Morphine**, named after the Greek god of dreams, was isolated in 1806. **Codeine** was discovered in 1832.[20] The problems of addiction to the analgesic (pain-relieving) natural opioids extracted from plants led to a search, around the time of the Second World War, for similar synthetic substances lacking their drawbacks.

By the early 1970s it had been shown that there were receptors for the opioids (naturally-occurring, or synthetic drugs with morphine-like actions) in the CNS of mammals, and that peptides with morphine-like effects existed in the brain. Research quickly demonstrated that there were, in fact, three types (families) of such opioid peptides. Peptides are short chains of linked amino-acids which are specified rather directly by the genetic material (DNA) of the cells producing them.[21] Each of the three families of opioids is derived from a polypeptide (long amino-acid chain) precursor. These are, respectively, proenkephalin, pro-opiomelanocortin (POMC) and prodynorphin.

As we have seen for other neurotransmitter substances, there are multiple types of receptor for these opioid peptides. The three main classes of receptor are termed μ (mu), κ (kappa) and δ (delta) and once again there appear to be many sub-categories of each. But as one standard source noted recently:[20]

> . . . It should be clear that our understanding of the detailed pharmacodynamic properties of opioid agonists and antagonists is in its infancy.

Agonist chemicals (drugs) have the same effects as a particular transmitter whilst antagonists have the opposite effects. An idea of possible future developments can be gathered from a consideration of a drug called **sufentanil**. Morphine binds about ten times as strongly to μ receptors as to δ and κ receptors, and it is thought to exert its analgesic (pain-relieving) properties mainly through its actions on μ receptors. Sufentanil appears to bind 200 times more tightly to μ receptors than to the other types.

Many semi-synthetic derivatives of morphine have been made and tested, for example: codeine (methylmorphine); heroin (diacetylmorphine); and apomorphine (the potent emetic). Etorphine, prepared from the related compound, thebaine, is a thousand times more potent than morphine. It is clearly possible, by altering their structure, to vary the effects of this class of chemicals on opioid receptors, as the greater affinity of sufentanil over morphine for μ receptors demonstrates. It is to be expected that such endeavours will become ever more systematic and subject to rational, computer-aided design.

Opioids such as morphine have many other effects on the CNS, one being the depression of breathing (partly by direct action on the respiratory control centres in the brain stem). Opioids can also be used to produce anaesthesia. **Fentanyl**, like the related sufentanil, is primarily a μ receptor agonist, but is only 80 times more potent than morphine as an analgesic. It is a synthetic opioid which is widely used as an anaesthetic, producing 'profound analgesia and unconsciousness' when large doses are administered slowly.[22]

Finally, it is perhaps interesting to note, remembering the original interest of the US Chemical Corps in the 1950s, that a rather similar series of discoveries to those concerning the structure and function of natural opioids has been made during recent research on tetrahydrocannabinol, the active constituent of marijuana. Receptors for this second class of chemicals have been found in the CNS, and a natural brain chemical, anandamide, with specific affinity for these receptors, has been extracted from brain tissue.[23]

Conclusion

This chapter has taken a brief look at an extremely complex subject. It would be difficult not to conclude that extremely rapid advances in our understanding of the structure and function of the human brain are likely over the next decade or two. This conclusion will be greatly reinforced in Chapter 7, where brain chemistry will be considered more from the perspective of the commercial drug companies. In that area of applied research, the driving force of potential medical benefit has combined with the attraction of huge financial rewards to produce an accelerating rate of change in our technological capabilities.

. 7 .

THE CHEMISTRY OF THE BRAIN

People have used drugs which affect the brain for thousands of years but only in the last half-century has systematic research produced a large number of new synthetic drugs. This is quite apparent if drugs which affect the central nervous system, and which are used in medical practice, are listed against their date of introduction (Table 7.1). Drugs which act on the CNS are among the most widely used of all medicines and, despite potentially dangerous side-effects, have brought great relief to many people. Unfortunately, because of their addictive properties, some of these drugs have been the cause of terrible personal and highly disruptive social problems. We still have much to learn about these substances.[1] Nevertheless, our developing knowledge is allowing an increasingly specific and systematic approach to prescribing, as is apparent from a classification of groups of drugs by their cellular targets (Table 7.2).

Rapid progress in scientific understanding and development of technological applications both seem possible in regard to receptors, particularly because of the link to modern genetic engineering. The point was made in an expert review of CNS-acting drugs:[2]

> With the ability to clone, sequence, and express the genes that encode receptor molecules for neurotransmitters, a new era in drug development is approaching. Such studies have permitted the identification of novel receptor subtypes that were undetected by traditional pharmacological approaches; the pace of such discovery will accelerate . . .

So it is with receptors that we should begin our analysis of the relevant chemistry of the CNS.

. 116

Table 7.1   Drugs that act on the CNS and which are used in medical practice*

Period of drug introduction	Indication
Before 1900	
Morphine	Pain
Caffeine	Drowsiness
Nitrous oxide	Surgery
Aspirin	Pain, fever, inflammation
1900 to 1950	
Barbiturates	Epilepsy
Phenytoin	Epilepsy
Meperidine (and analogues)	Pain
Antihistamines	Wakefulness
Since 1950 (examples only)	
Lithium carbonate	Bipolar affective disorders
Chlorpromazine (and related phenothiazines)	Psychoses
Haloperidol (and related butyrophenones)	Psychoses
Monoamine oxidase inhibitors	Depression
Tricyclic antidepressants	Depression
Mixed agonist/antagonist opioids	Pain
Methadone	Opiate dependence
Clonidine (and related imidazolines)	Hypertension
Diazepam (and related benzodiazepines)	Anxiety
L-DOPA	Parkinson's disease

* Data from reference 1

Receptors

Neurons are cells specialised for processing information. The neuron transmits information by means of action potentials travelling down the axon. Each axon potential is of the characteristic size for that axon and proceeds from its initiation point to its axon terminals. The information carried is encoded in the frequency of the action potentials. The process is fundamentally unitary. The more excitatory input the neuron receives, the greater the frequency of the action potentials it generates.

Table 7.2 Some drug targets in the CNS*

Cellular target	Class of drug
Opioid receptors	Opioid analgesics
Opioid receptors	Opioid antagonists
GABA receptors	Benzodiazepines
GABA receptors	Barbiturates
Adrenergic receptors	Clonidine
Histamine receptors	Antihistamines
Dopamine receptors	Antipsychotic drugs
Norepinephrine reuptake	Tricyclic antidepressants
Serotonin reuptake	Tricyclic antidepressants

* From reference 1

What then happens when an action potential reaches the region where the axon synapses onto a post-synaptic nerve cell? As we saw in Figure 5.1 at B, the neurotransmitter in the pre-synaptic cell is characteristically stored in small vesicles near the synaptic cleft. As with the action potential in the axon, the process at the presynaptic terminal is fundamentally unitary. Unless other events intervene, such as depletion of the transmitter substance in the presynaptic terminal, a similar amount of transmitter is released by each arriving action potential. In addition to the frequency of the arriving action potentials, what determines the post-synaptic response is the receptor in the post-synaptic cell membrane.

Though what happens at synapses is complex, one point is apparent. The process at the post-synaptic membrane is not often the unitary passing on of one action potential for each incoming action potential. Further levels of complexity may be understood from Figure 7.1A. The transmitter substance could have other potential targets in addition to the immediate post-synaptic receptor. It could also perhaps affect **auto-receptors** located near the synapse, on the pre-synaptic cell. If these are inhibitory in effect, transmitter release will tend to be reduced by its own release, a process of auto-feedback. The post-synaptic cell could also have synapses back onto the pre-synaptic cell, as could other cells in the region where the transmitter is released. Their feedback effects on the pre-synaptic cell could be either positive or negative. Moreover, all the cells in the region are likely to be incorporated in a variety of

Figure 7.1   Synaptic Transmitters and Receptors

A. Transmitter Destinations

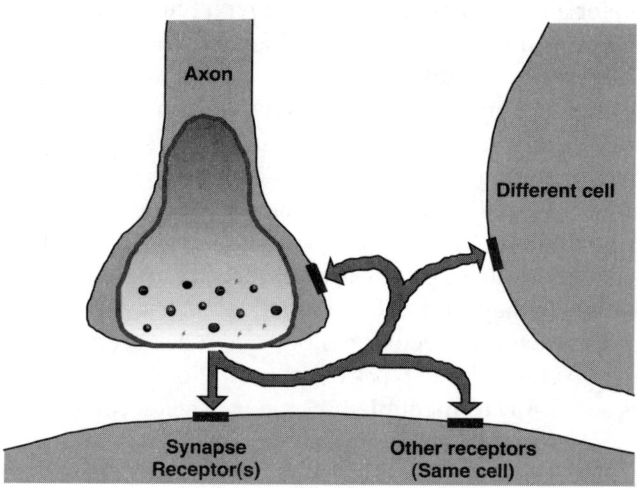

B. An Alpha₂ Adrenoceptor

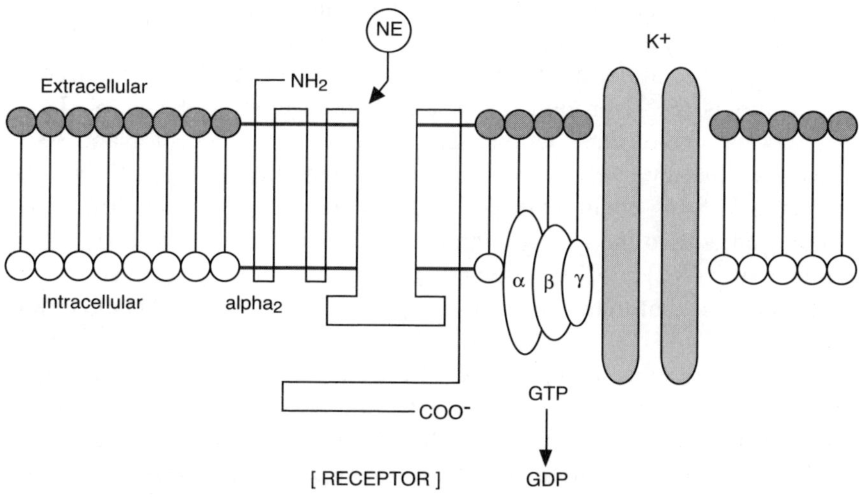

*After references 1 and 6

different neuronal networks. So it may be seen that in simply administering a drug such as morphine or BZ, even if it produces a clear-cut, observable effect on behaviour, we are a long way from being able to say how that effect is achieved. Nevertheless, knowledge is growing.

A receptor is defined as 'a binding site with functional correlates' and the natural transmitter is called the **endogenous ligand**.[3] Receptors can be classified in several different ways such as anatomically, according to where they are found (eg in the sympathetic nervous system), or pharmacologically, according to the transmitter substances that most affect their operation. Table 7.3 provides a recent standard listing of receptors in mammalian species. As noted previously, numerous subtypes of the different kinds of receptor have also been identified.

The most interesting classification of receptors is based on the mechanisms by which they work.[4] This classification has been made possible by the application of molecular biology and it has been discovered that there are, in fact, two basic types of receptor: **ligand-gated ion channel** receptors and **G-protein-coupled** receptors. The receptors, importantly, are composed of proteins. The opiate receptors discussed at the end of the last chapter are of the second type, which involves a G-protein.[3,5] Because of what follows in Chapter 8 we shall briefly discuss here the **alpha-2 (α_2) adrenoceptor** (adrenergic receptor) which is also a G-protein-coupled receptor.[6]

Each G-protein-coupled receptor has only one protein unit. This crosses the cell membrane seven times (line from NH_2 to COO in Figure 7.1 at B). The important point is that associated with the receptor protein unit is the G-protein, made up of three units α, β and γ. This functions as the transduction element between the receptor and secondary messenger systems inside the post-synaptic cell which ultimately bring about its response. In the absence of transmitter, the β and γ units hold the α unit in an inactivated state. However, when the transmitter (NE) attaches to the receptor protein there is a change which allows the α unit to become active. The action of the α unit changes an enzyme called GTP to GDP. In the alpha-2 (α_2) adrenoceptor the G-protein is described as inhibitory (G_i) because the events activated by the secondary messenger (including potassium [K^+] ion movements out of the cell) are inhibitory in effect. The inhibitory function of the alpha-2 adrenoceptor is of particular interest because the receptor is frequently found pre-synaptically as an inhibitory auto-receptor on NE transmitter-producing cells.

That is by no means the only complexity when attempting to understand the operation of alpha-2 adrenoceptors. We do not need to go into

Table 7.3   Recent listing of receptors*

Adenosine (P_1 purinoceptors)

α_1-Adrenoceptors

α_2-Adrenoceptors

β-Adrenoceptors

Angiotensin

Atrial natriuretic peptide

Bombesin

Bradykinin

Calcitonin gene-related peptide and amylin

Cannabinoid

α-Chemokine

β-Chemokine

Cholecystokinin

Dopamine

Endothelin

Excitatory amino acids

GABA

Galanin

Glycine

Histamine

5-Hydroxytryptamine

Leukotriene

Melatonin

Muscarinic acetylcholine

Neuropeptide Y

Neurotensin

Nicotinic acetylcholine

Opioid

Platelet-activating factor

Prostanoid

P_2 purinoceptors

Somatostatin

Tachykinin (Substance P is an endogenous ligand)

Vasopressin and oxytocin

Vasoactive intestinal and related peptides

---

* From reference 3

the details which may underlie processes such as desensitisation to drugs here, but it is necessary to get an overview of the catecholamine systems of the CNS before examining the broad consequences of the effects of the major groups of drugs on the functioning of the whole CNS.

Catecholamines are organic compounds with a catechol nucleus, ie a benzene ring with two adjacently-placed hydroxyl (-OH) groups, and an amino group ($-NH_2$) attached.[7] Examples we have discussed are nor-epinephrine (NE, noradrenaline), epinephrine (adrenalin) and dopamine (DOPA). All these compounds are linked in a simple and well-known synthetic pathway. Starting with the amino-acid tyrosine, which is found in many foods, a series of cellular enzymes can produce, first, L-DOPA, then dopamine, norepinephrine and finally epinephrine. Neurons with all the relevant enzymes can produce epinephrine, those with the first two can produce dopamine and those with the first three can produce norepinephrine. Table 7.4 sets out the series of stages by which NE, produced from dopamine, is used as a transmitter. The Table also indicates how various drugs may be used to interfere with or potentiate these stages. NE differs from acetylcholine (ACh) in that it is not destroyed by an enzyme in the synaptic cleft, but is reabsorbed (so-called reuptake) into the presynaptic cell. However, free NE in the cell can be degraded by the enzyme monoamine oxidase (MAO).

Table 7.4   Possible sites of drug action on an NE (norepinephrine) neuron*

Stage	Drug
1. Synthesis of NE from DOPA	Blocked by Fla-63
2. Storage of NE in vesicles	Reserpine interferes with the uptake/storage mechanism
3. Release and effect of NE on receptor	Clonidine has similar (agonist) and yohimbine (antagonist) effects at α_2 receptors
4. Reuptake from synaptic cleft	Inhibited by amphetamine
5. Destruction of free NE in presynaptic cell	Monoamine oxidase (MAO) action inhibited by iproniazid

* From reference 7

Despite the individuality of different transmitter receptor systems, it is known that drugs which have been developed in an attempt to target various specific sites (Table 7.2), when administered globally to the whole body, have the potential to affect many different body systems. It is necessary to consider the current knowledge of such global drug effects in the CNS.

The Action of Drugs on the Central Nervous System

It would be convenient if we could classify drugs which affect the CNS by their mechanism of action but, since this is not presently possible, other means must be sought. A second method of systematisation might be to classify drugs by their chemical structure but, as Julien has indicated:[8]

> . . . too many drugs with similar structures induce different pharmacological activity, and too many drugs with dissimilar chemical structures induce pharmacological activity that is nearly identical . . .

So this approach is unhelpful too. Despite the problems caused by the diverse potential effects of a particular chemical, the most useful categorisations remain pragmatic ones based either on the most characteristic behavioural effects or on major clinical use of a particular drug. Julien's own classification, made on the basis of these two criteria, is set out in a simplified form in Table 7.5.

Table 7.5   Simplified classification of drugs affecting the CNS*

1. Traditional, non-selective CNS depressants
2. Antianxiety agents
3. Antiepileptic drugs
4. Psychomotor stimulants (psychostimulants)
5. Antidepressants
6. Mood stabilisers
7. Narcotic analgesics: opioids
8. Antipsychotic agents
9. Psychedelics and hallucinogens

* From reference 8

CNS depressants

The problem with Julien's approach is obvious from a consideration of **Group 1** in his classification: CNS depressants. He notes that:[8]

> The terms sedative, tranquillizer, anxiolytic and hypnotic can be applied to any CNS depressant because each may diminish environmental awareness, spontaneity, and physical activity; higher doses produce drowsiness and lethargy; and even higher doses produce sleep . . .

We are clearly dealing with varying degrees of depression of the CNS and the chemicals which produce these effects may be very different indeed. Moreover, depending on dose, the same chemical may produce effects ranging from sedation to anaesthesia. Julien lists four types of chemical in Group 1: the barbiturates; the non-barbiturate hypnotics; ethyl alcohol; and the benzodiazepines. He also lists one functional category – general anaesthetics – which is of particular interest to us. The use of **ethyl alcohol** has, of course, an even longer history than that of the opioids but it was not until the middle of the last century that new means of inducing sedation, such as the non-barbiturate, **chloral hydrate**, became available. **Phenobarbital** was introduced in the early years of this century and some 50 other barbiturates came on to the market before 1950. Only in 1960 did **chlordiazepoxide** (Librium), the first benzodiazepine tranquillizer, reach the market.

Though the details are complex, basically both barbiturates and benzodiazepines alter the $GABA_A$ receptor (GABA or γ-amino butyric acid is a natural inhibitory transmitter which depresses neuronal activity). The two groups of drugs have been widely used in the treatment of sleep disorders, anxiety states (Julien's **Group 2**) and epilepsy (**Group 3**, due to their reduction of neuronal activity),[1] but there has been increased caution in prescribing them in recent years because of side effects which include the potential for addiction.[9]

General anaesthesia is the most severe state of CNS depression which combines:[1]

> . . . analgesia, amnesia, loss of consciousness (absence of awareness), relaxation of skeletal muscles, suppression of somatic, autonomic and endocrine reflexes, and hemodynamic stability . . .

There are two means of achieving this state: by inhalation of various gases or by intravenous injection of various drugs. The mechanism of action of the injected anaesthetics is generally accepted. The barbiturates and benzodiazepines act on the $GABA_A$ receptor.[1,10] Other intravenously injected drugs such as opioids have different mechanisms of action:[1]

. . . The depressant effects of morphine-like opioids on neuronal activity are mediated by the μ-opioid receptor. . . . Ketamine appears to act not by enhancing neuronal inhibition but by blocking neuronal excitation . . .

Ketamine has this effect, it appears, through binding to the phencyclidine site on the NMDA glutamate receptor and thereby blocking the excitatory action of glutamic acid.[11] **Clonidine**, on the other hand, exerts its action through the alpha-2 (α_2) adrenergic receptor. It may be noted here that:[10]

. . . Knowledge of the site and mechanism of action of the α_2-adrenergic agonists may provide the selectivity and specificity that is so often lacking in other drugs that are used in anaesthetic practice . . .

We shall take up this point again in the next chapter when discussing the US Army's interest in dexmedetomidine.

The mechanism of action of inhalation anaesthetics is generally much less well understood.[1] As Julien noted:[8]

Prominent scientists have studied the possible mechanisms of action of the inhaled anesthetics for over 100 years. The mechanisms proposed have been so diverse that one recent textbook of pharmacology devoted seven pages to this puzzle . . .

However, a recent major review in the journal *Nature* argued that, whilst there were clearly exceptions, most general anaesthetics exerted their effect predominantly through action on the $GABA_A$ receptor, as we have described above for the barbiturates and the benzodiazepines.[12]

CNS stimulants (psychomotor stimulants)

In **Group 4** of his classification of drugs which affect the CNS Julien includes **caffeine** (adenosine receptor blocker), **nicotine** (acetylcholine receptor agonist), **cocaine** and the **amphetamines**. Clearly, we are again dealing with a diverse set of drugs which in moderate doses increase arousal and reduce fatigue. The reason for the alternative names in the sub-heading is that:[9]

. . . As many drugs that produce such actions also increase electrical activity in the CNS and in the sympathetic branch of the peripheral nervous system, they are sometimes termed CNS *stimulants* or *sympathomimetic agents* . . . . *psychomotor stimulants* emphasises the fact that agents so categorised may elevate motor activity.

Cocaine occurs in the leaves of the coca plant, indigenous to parts of South America, and has been used by local populations there for more than 1,500 years.[13] Cocaine was chemically isolated in the middle of the last century and became widely used as a patent medicine. Misuse of cocaine declined as amphetamines became popular during the 1930s, but the availability of high-dose, rapid-onset crack cocaine has led to widespread abuse since the 1970s.

Cocaine causes 'a grossly exaggerated mobilisation of our normal fight/flight/fright response', which is dependent on the effects of the drug in increasing the action of dopamine and norepinephrine.[8] The mechanism involved is thought to be:[13]

> . . . An increase in CNS dopamine neurotransmission, resulting from a competitive blockade of high-affinity dopamine uptake . . .

Neither norephinephrine nor dopamine is degraded by enzymes within the synaptic cleft. Termination of their post-synaptic effects is therefore dependent on their being actively reuptaken into the presynaptic nerve terminals. Cocaine, in blocking this reuptake, allows excessive amounts of dopamine in particular to persist and remain active. Increasing or potentiating the effect of normal dopamine output in this way produces the pronounced behavioural effects of the drug, although the precise mechanisms and neuronal circuits involved are not yet identified.

Amphetamine was synthesised in the latter part of the 19th century but did not begin to be used medicinally until the 1930s. The number of conditions for which it was prescribed then rapidly increased and in the Second World War it was widely used as a means of counteracting fatigue in the armed forces. After the war there was increasing abuse of amphetamines.[9] Amphetamines appear to cause their pronounced behavioural effects through:[8]

> . . . the release of newly synthesised catecholamines (especially dopamine) from presynaptic storage sites in nerve terminals . . .

The overall effect, therefore, is similar to that of cocaine because, 'both drugs have the net effect of increasing the amount of dopamine available'. Again, the mechanisms whereby the drug produces behavioural effects are complex and not yet well understood.

Antidepressants

The drugs in **Group 5** of Table 7.5 have been used to treat the most common mental problem in adults, depression. Standard terminology today differentiates between unipolar depression and bipolar illness (previously termed manic-depressive illness). **Unipolar depression** affects between six and ten per cent of the population at some time during their lives.[1] Obviously, most people experience sadness and mild depression from time to time but we are referring here to a much more severe change in mood. The exact causes of the illness and the reasons why drugs in Group 5 actually work are subject to some debate.[14] However, the current predominant view is that 'depression may involve deficits in brain biogenic amines',[1] which leads to the possibility that:

> . . . increased catecholamine function is one mechanism by which antidepressant drugs may alleviate depression.

This certainly appears to be what happens when all three major types of antidepressant drug are used. For example, **tricyclic drugs** block the reuptake of transmitters such as norepinephrine and serotonin. **Monoamine oxidase inhibitors**, on the other hand, increase the amounts of the amines available for release at the presynaptic nerve terminal by blocking the action of the enzyme monoamine oxidase (MAO), which normally breaks down these transmitter substances. One problem with the idea of a simple connection between the use of these drugs to increase the amount of available transmitters and the lifting of depression is that transmitter levels rise quickly after a course of drugs is begun, but several weeks elapse before a change of mood is apparent.

The final group of antidepressants – the **second-generation, atypical antidepressants** – was produced in the late 1970s to mid-1980s in an effort to overcome the limitations of previously available drugs. They include the serotonin-specific reuptake inhibitors such as **fluoxetine** (Prozac) which has now achieved the public popularity previously held by drugs such as Librium and Valium. Yet despite much medical effort, the fact remains that 20 per cent of people with depression do not respond to treatment or cannot tolerate the side-effects of the drugs available.[8]

Mood stabilisers

Bipolar illness may involve depression, but is characterised by one or more manic periods in which the person has little need for sleep, talks a great deal, has racing thoughts and so on. Although the incidence of the disease in the population is low (up to one per cent), it does seem to have a genetic component, tends to recur and can be extremely disabling. The drug which is most successful in treating this disease is **lithium**. Other mood stabilisers in **Group 6** of Table 7.5 are carbamazepine and val-proate which are used if lithium treatment fails. Lithium is an alkali metal and its prescription is successful in some 60–80 per cent of cases.

It has been used previously for other medical purposes but its effectiveness in treating bipolar illness was discovered only in the late 1940s. In the concentrations used, lithium has no discernible effect on a healthy person and is of little use in lifting a depressive phase of the illness. The mechanism by which lithium exerts its spectacularly successful effects on the manic phase is not yet known. Wide-ranging suggestions have been made, frequently involving amine transmitters, but as yet there appears to be no clear consensus among research workers.[8]

Opioids

We have already briefly discussed the **Group 7** drugs, the narcotic analgesics or opioids. **Morphine** exerts its well-known pain-relief and other effects primarily by its action on μ (mu) receptors, which are located mainly above the level of the spinal cord:[8]

> . . . Such areas include the medial thalamus and brainstem areas . . . . Activation of mu receptors in these areas seems to underlie the analgesic, respiratory depressant, miotic (pinpoint eyeballs), euphoric, and physical dependence properties of morphine and related drugs . . .

At present, although it is thought that respiratory depression is linked to a specific receptor sub-type, it is not possible to devise a drug which produces analgesia without also depressing respiration.

The opioids which occur naturally in the human body, the **endorphins**, in part function in a positive reward/feedback system which has evolved to reinforce actions such as eating which are necessary for survival. The ability to use other opioids, made outside the body, gives us direct access to that powerful reward system. So when morphine is used, or **heroin**, there is not only a feeling of euphoria, but a powerful reinforcement of the action. Considerable research has produced detailed proposals on the mechanisms underlying the acquisition, maintenance and relapse of opiate addiction,[15] and on how the reward pathway might be related to those underlying addiction.[16] Again, there appears at present to be no clear consensus on the precise mechanisms involved.

Antipsychotic agents

Schizophrenia is one of the most severe mental illnesses and affects about one per cent of the human population worldwide. The symptoms of the illness are generally described as being of two distinct types:[8]

> . . . The positive symptoms are those typical of psychosis and include delusions and hallucinations, bizarre behaviors, dissociated or fragmented thoughts, incoherence, and illogicality . . .

On the other hand:

> . . . The negative symptoms include blunted affect, impaired emotional responsiveness, apathy, loss of motivation and interest, and social withdrawal . . .

Schizophrenia clearly has a genetic component because there is an increased risk of developing the illness amongst close relatives of sufferers.[1] The disease strikes when people are young and is progressive. The costs to the community of caring for the many people with the illness are therefore considerable and, until the 1950s, there was little that could be done to treat them.

Of the antipsychotic agents, Julien's **Group 8, chlorpromazine** was the first to be used to treat mental illness, in the 1950s.[9] The effectiveness of treating schizophrenia with such drugs has been established by numerous carefully controlled trials. Chlorpromazine is a member of the **phenothiazine** family of agents. Other phenothiazines have similar effects to chlorpromazine in successfully relieving the symptoms of schizophrenia. Other classes of drugs such as the **butyrophenones** (eg haloperidol) have been developed since the initial successful drug treatment of the illness. It appears that all these drugs have an affinity for, and block, a specific type (D_2) of dopamine receptor in the brain and:[1]

> . . . Multiple lines of investigation indicate that the unifying principle of antipsychotic action of these drugs may be the reduction of dopamine synaptic activity in the limbic forebrain . . .

Yet again, unfortunately, the situation is very much more complex than this simple theory would imply. It appears that a multifactorial theory will have to be developed to account fully for the illness and its relief.

Psychedelics

The psychedelic drugs in Julien's **Group 9** include natural substances which have been used for various, often religious ritual, purposes by human groups for millennia. In the modern Western world there was little awareness of such substances but, since the Second World War, interest has grown and newer agents have been developed in legal and illegal[17] laboratories. Julien makes the point that the chemical structures of this group are very diverse, but he subdivides them into five classes: **anticholinergic**; **catecholamine-like** (ie norepinephrine-like); **serotonin-like**; **psychedelic anaesthetics**; and **tetrahydrocannabinol** (Table 7.6). This classification is based mainly on the types of neurotransmitter which the psychedelics affect. As Julien notes:[8]

. . . The anticholinergic psychedelics produce intoxication, amnesia, and delirium by blocking postsynaptic acetylcholine receptors. The catecholamine-like and serotonin-like psychedelics act by influencing a subset of serotonin receptors . . .

The psychedelic anaesthetics such as **ketamine** have a complex action on an excitatory glutamate receptor, as noted previously.

The active ingredient of marijuana, **tetrahydrocannabinol**, as mentioned in the last chapter, has specific receptors in the brain. However, Julien considers it to be a different category of agent altogether from the other four groups, one requiring separate treatment.

Table 7.6   Classes of psychedelic drug*

---

1. **Anticholinergics**

   – Atropine
   – Scopolamine

2. **Catecholamine-like**

   – Mescaline
   – MDMA (Ecstasy)

3. **Serotonin-like**

   – LSD
   – Psilocybin

4. **Psychedelic anaesthetics**

   – Phencyclidine
   – Ketamine

5. **Tetrahydrocannabinol**

   – Marijuana

---

* From reference 8

A comparison of the agents mentioned in Table 7.6 with the history of non-lethal agents in Chapter 5 shows that one of the groups most interested in increasing its knowledge of these psychedelic substances and their effects was the military. We should therefore examine some of the potential agents individually before considering, in Chapter 8, the US Army's continuing search for non-lethal incapacitants.

The **anticholinergics** scopolamine and atropine are found in high concentrations in the plant *Atropa belladonna* (Deadly Nightshade) which was used in the past as a source of poison. Both substances prevent acetylcholine action by blocking its receptor sites. However, atropine crosses the blood-brain barrier much less easily than scopolamine so it is the latter which is of most interest here. Primarily, the effect of the drug is not to 'expand consciousness' but to cause confusion, sedation and amnesia.

The **catecholamine-like** and **serotonin-like** psychedelics are said to produce a very similar set of effects, including:[8]

> . . . sensory-perceptual (time) distortion; altered perceptions of colors, sounds, and shapes; complex hallucinations and synesthesia; dream-like feelings; depersonalisation; altered effect (depression or elation); and somatic effects (tingling skin, weakness, tremor, and so on) . . .

It appears that the drugs in these two subgroups are partial agonists for one type of serotonin (5-hydroxytryptamine) receptor and their effect is to set off a cascade of events of great complexity within the central nervous system. **Mescaline** is derived from a cactus common in the south-west of the United States and is still used legally in religious practices by American Indian tribes. In common with other catecholamine-like psychedelics, it bears some resemblance to amphetamine, but has additional chemical features which, although not complex, appear to confer the extra psychedelic properties. Many related compounds have been synthesised on the basis of this chemical knowledge – MDMA or 'Ecstasy' being one example. The serotonin-like psychedelics include **lysergic acid diethylamide** (LSD) as well as the mushroom-derived **psilocybin** and **psilocin**. LSD was synthesised in the late 1930s and its remarkable properties were discovered by accidental ingestion. As we saw in Chapter 5, LSD was the subject of extensive research during the 1950s, and then spread from college campus experimental volunteer use to widespread popularity among the general public in the 1960s. It is effective in very small amounts and has low toxicity:[8]

> . . . Gable estimates for LSD an effective dose of 50 micrograms and a lethal dose of 14,000 micrograms. These figures provide a therapeutic [safety] ratio of 280, making the drug a remarkably non-lethal compound . . .

However, the psychological responses to LSD are intense, highly variable, and can be very dangerous for the individual involved. So despite its low effective dose and low toxicity, LSD would not be a useful incapacitant agent. On the other hand, it is not addictive.

Psychedelic anaesthetics such as **phencyclidine** ('angel dust' or PCP, a popular illicit drug in the 1970s) and **ketamine** are not structurally related to the drugs just mentioned. Phencyclidine was first produced in the mid-1950s and initially was used as an anaesthetic for humans. Its use was abandoned, however, because of many serious psychotic reactions. Ketamine produces such side effects much less frequently and is still used as a human anaesthetic in special circumstances. Phencyclidine has been found to act specifically on the post-synaptic NMDA glutamate receptor. Ketamine has a similar mode of action, as described earlier in the chapter in the discussion of anaesthesia. When used for anaesthetic purposes these drugs are called dissociative anaesthetics because they produce an unusual response:[8]

> . . . They induce an unresponsive state with intense analgesia and amnesia, although the patient's eyes remain open (blank stare), and he or she may even appear to be awake.

When used illicitly, the effects may be highly variable and dangerous. Phencyclidine seems to be the only one of the psychedelic drugs that is addictive.

Tetrahydrocannabinol (THC) is the active constituent of the hemp plant *Cannabis sativa*. Active preparations from this plant have long been widely used and are still used under names such as marijuana, cannabis and hashish. In moderate doses the drug has a sedative-hypnotic effect, but in higher doses, unusually for a drug with such depressive properties, it can produce euphoria, hallucinations and so on. So this drug combines the properties of both sedatives and psychedelics. It is safe in high doses and, as we saw in the last chapter, it is also now thought to have a natural analogue and specific receptors in the brain. Despite vigorous attempts to outlaw the use of marijuana in the Western world, it became widely popular in the 1960s and 70s and is still so today. Recent studies have shown that the THC receptors are located in the basal ganglia and cerebellum, the cerebral cortex (especially the frontal cortex) and the hippocampus. This distribution would seem to account for the known effects of the drug on movement control, psychedelic experiences and memory respectively. There do not appear to be receptor binding sites for THC in the brain stem and this in turn would account for its non-interference with essential regulatory functions and thus its non-lethality. However, long-term use of marijuana does appear to have a side-effect – suppression of the immune system.

Mechanistic neuroscience

In conclusion then, it is not difficult to see why neuroscientists are so excited over the discoveries which may be made in the near future about the functioning of the human brain. An enormous base of knowledge is already available and a variety of modern techniques have now been developed which, particularly in combination with molecular biology, are likely to increase rapidly our understanding of how drugs with diverse structures affect different receptor types.[18] Such a mechanistic understanding will increase our power to provide medical help to those in need but also, perhaps, our capability to wage a new form of warfare. Before returning to that issue, a brief consideration of international efforts to control illegal drugs will illustrate the problems we may have in the future in trying to erect arms control treaties to control non-lethal weapons.

Social Control of Drug Abuse

Substances which are subject to abuse may be defined as:[1]

> . . . drugs or other materials (eg solvents) administered repeatedly in
> a pattern and amount that interferes with the health or normal social
> and occupational functioning of the individual.

Psychological dependence on such drugs can involve 'intense craving' and 'compulsive drug-seeking' and, as noted previously, drugs which are subject to abuse induce very strong reinforcing mechanisms involving 'intense euphoria and feelings of well-being'. For the individual taking the drug, and for the medical professional trying to treat the drug-taker, this combination of effects presents an almost intractable problem, despite the known dangers of drug abuse. However, the problem is more than an individual one. What must also be considered are some of the problems of social control related to drug abuse, particularly international attempts to deal with the trading in illegal drugs by criminal groups.

The social costs of illegal drug-taking are well known, but worth emphasising:[19]

> Over the past decade, the US government has spent $53 billion
> enforcing drug laws, locking up 260,000 people . . . in the process.
> And for what? In inner cities, crack and heroin use stubbornly refuse
> to shrink . . . while in some American prisons, drug offenders account
> for more than half the inmates.

And the social costs go well beyond this:

> . . . In 1993 almost £2 billion worth of drug-related theft – about
> £114 per household – was committed in Britain . . .

The situation is so bad that some of those directly involved in tackling the problem argue that it would perhaps be better to give up the national domestic fight and legalise at least some of the drugs.

In an address to the House Foreign Affairs Committee in June 1994 the US Assistant Secretary for International Narcotics Matters stressed the international aspect of control:[20]

> ... The effects on American society if we fail to address the narcotics problem abroad will be direct and unambiguous: more addiction, crime, violence, disease and poverty. ... We need greater international co-operation to overcome this threat ...

As the then Assistant Legal Advisor to the US Department of State pointed out in 1990,[21] there is in fact a long history of international co-operation on this issue (Table 7.7). Unfortunately, the arrangements in the 1970s proved inadequate:

> ... In the following decade, as the power of the drug cartels became more pervasive and their methods increasingly sophisticated, the need for new and more stringent international measures became clear ...

Thus within the United Nations organisation, work was undertaken from which evolved a new UN Convention Against Illicit Traffic in Narcotic Drugs and Psychotropic Substances in 1988.

Table 7.7 International agreements on drug control*

1912	**International Opium Convention**
1961	**Single Convention on Narcotic Drugs** (consolidated earlier agreements)
1972	**Protocol to the Single Convention** (strengthened provisions against illicit narcotics)
1971	**Psychotropic Substances Convention** (extended control to man-made hallucinogens, stimulants and sedatives)

* From reference 21

This Convention not only calls upon parties to enact domestic legislation to control illicit trafficking and money-laundering, but also to back this up by making offences the basis for international extradition and for mutual legal assistance between states. States are also obliged to co-operate in suppressing illicit traffic and in eradicating illicitly grown crops. Finally, the Convention is strengthened by having its administration and oversight in the hands of the UN Commission on Narcotic Drugs and the UN International Narcotics Control Board.

Despite the well-known difficulties of co-ordination between the legal systems of different states:

> . . . the Convention requires party states to enable their courts or other competent authorities to order the production or seizure of bank, financial or commercial records necessary to trace, identify, seize and forfeit proceeds and instrumentalities of drug trafficking . . .

Furthermore, it is intended to be made impossible to decline to do this on the grounds of bank secrecy legislation. At first sight, this might appear a great success for international collaboration in dealing with a major transnational problem, but critics argue that the Convention on illegal trafficking is fundamentally weak, and matters are made worse by the US Senate 'placing reservations on some of the very items the US negotiators had fought very hard to get' during ratification.[22] Others argue that the international community will never be able to deal with the problem of illegal drugs unless it recognises the links to conditions in poor and conflict-prone regions of the world where the raw materials may be produced, and to the unregulated nature of the world financial system which facilitates laundering of drugs-related profits.[23] What is clear is that the international drugs trade and the estimated annual revenues of some $150-$200 billion derived from it are not remotely near being brought under inter-governmental control.

.8.

AN ASSAULT ON THE BRAIN?

The use of CS gas by US forces in Vietnam was discussed in some detail in Chapter 5, as was the development of the incapacitants programme which eventually produced BZ as an interim standard agent in the early 1960s. Appendix 2 gives an insight into the understanding that the US military had of the chemistry of the nervous system when they began their work. It will be recalled that, at that time and given the progress of medical research, there was 'a good deal of optimism regarding the possibility of fulfilling military requirements'. We should now be in a better position to examine whether this optimism is still felt in military circles today.

It is important to note that, despite the rapid progress made in recent years towards achieving a comprehensive Chemical Weapons Convention, there really is considerable doubt in the military over the assumption that chemical weapons will never be used again, or that they will only ever be used in actions which do not involve Western powers.[1,2] There have certainly been alarming suggestions about the possible development of incapacitating chemical weapons among potential adversary states,[3] and there have also been serious military studies of such possibilities.

The Foreign Science and Technology Center of the US Army Intelligence Agency, for example, produced a report in 1986 with the title, 'Incapacitating Agents, European Communist Countries: Neurotoxins, Psychotropics, and Organofluorines'. The report stated that:[4]

The current mission of the 476th Military Intelligence Detachment (Strategic) is to investigate Soviet research, development and use of incapacitating agents by constant and careful analysis. . . . Based on this analysis, the nature of the threat posed to our military forces by incapacitating agents at the present time is assessed, and finally a forecast is made as to what forms the threat may take in the future.

The report was based on the study of a very large amount of data. One of its conclusions was that:

> The Soviets seem to be concentrating their largest effort in studies of the cholinergics, LSD and phenothiazines. Their synthesis of various compounds lends credence to their interest in research toward incapacitating type agents that may be used as powerful psychochemicals with potential chemical warfare applications.

It was argued, furthermore, that this research interest dated at least from the 1960s. The subject will not be pursued here, save to emphasise that though what follows mainly describes work in the United States in recent decades, it would be foolish to imagine that such work could not have been mirrored in other countries.

Over the years since the Vietnam War, US interest in incapacitating agents has been reported occasionally in the non-technical literature. The 1982 *SIPRI Yearbook* reported that a number of agents had been standardised and then abandoned because of various shortcomings. According to SIPRI, these abandoned agents included **phencyclidine** (agent SN) and 3-quinuclidinyl benzilate (**BZ**). The authors argued that:[5]

> . . .The search continues. As of 1977 the leading contender was something known in public only as EA 5302, which is a solution of a psychotropic glycollate related to BZ (EA 3834B) in a novel volatile irritant liquid coded EA 4923, apparently a cycloheptatriene.

In 1984 the *SIPRI Yearbook* reported that the status of the incapacitating chemicals programme in the USA was obscure.[6] All existing stocks of BZ and BZ-filled munitions were due to be destroyed and, in 1981, US Army research and development planning had envisaged a range of new munitions replacing these systems in the late 1980s. However, no explicit funding for this purpose was evident in the 1984 budget. The authors were able to report, though, the chemical identities of agents **EA 3834** and **EA 4923** and to say something about their properties. EA 3834, the candidate agent which had been expected to replace BZ in the 1970s, was stated to have a median incapacitating airborne dose by inhalation of about half that of BZ. It was also said to have a strong percutaneous (skin-permeating) activity when in solution. This last characteristic has obvious military importance since it would require protection through the use of a full protective suit rather than just a respirator.

The quest for such agents obviously continues. In 1993 the 'Small Business Innovative Research Program' of the US Department of Defense advertised for work to be carried out on 'less-than-lethal immobilising chemicals'. The aim of the project was:[7]

> To suggest, acquire, evaluate and develop chemical immobilising material for application to various missions such as: rescue, embassy protection, anti terrorism, barricade situations, domestic disturbances, and other law enforcement scenarios.

The advertisement explained that recent programmes under this title at the US Army Edgewood Research, Development and Engineering Center had focused on **fentanyls** as potential agents. However, though these were well-characterised, rapid, potent and reliable, they did not have high enough safety ratios between effective and lethal doses and their duration of action was too long. Thus other candidate immobilisers without such drawbacks were still required.

Recently, the author of many of the SIPRI studies, Dr Julian Perry Robinson of the University of Sussex, has provided an overview[8] and a chronology[9] of the official information presently publicly available on the US (and earlier UK) efforts to produce an effective incapacitant. A brief consideration of these important studies will enable current known activities to be put into proper perspective. We can then assess what may happen in the future in the light of what we now also know of current open research in medicine and biology. Perry Robinson's detailed chronology covers all aspects of the subject, from the command decisions and organisations involved in the effort to produce incapacitating munitions over the five decades from 1945, through to the technical characteristics of the chemicals themselves. For present purposes it is sufficient to concentrate on the nature of the research and development programme: what was being attempted and when. We need to know what chemicals have come close to being the replacement for BZ and why. With that information we should be able to set what was done in the context of the total research on the chemistry of the human nervous system. Certain historical studies, originally carried out for governmental purposes, have become available for public scrutiny and are, as we saw in Chapter 5 (and Appendix 2), particularly helpful for our purposes. We can also use certain reports on the American programme which were regularly made to Congress through most of the latter half of this 50-year research and development programme.

The Search for Incapacitants

For simplicity we shall deal with the information chronologically.

The early years

An insight can be gained into the search for incapacitants by reviewing a study carried out by the US Army, in the mid-1970s, of its chemical agent testing between 1950 and 1975, ie the first half of the 50-year research and development programme. The need for this study arose from increasing public interest in just what testing had been done on humans – including testing without informed consent – by the intelligence community in the United States during this period. The Army found that it was being asked questions about its involvement without having an adequate historical record on which to base its replies. A detailed study was therefore authorised in order to put the necessary data together.

As the report makes clear, this was no easy task because the period in question:[10]

> . . . probably had as many changes, programs, and problems as any comparable period in history: post-World War II; the Korean war; the Cold War; reorganisation of the . . . Army; the war in Vietnam; and major advances in medicine, the sciences . . .

There were also problems with the quantity of data to be analysed, and because some of it had been routinely destroyed. Nevertheless, an idea can be gained of what was involved in the search for incapacitant agents from a flow chart produced in the report (Figure 8.1). In a wide-ranging search, potential chemical agents (irritants, incapacitants or lethals) were gathered from a variety of sources. If an agent was thought to be a possible incapacitant, it then entered the process set out in the Figure. An initial biological evaluation was made of its effectiveness. Any compound passing this screening was subjected to a much more complex series of secondary tests. Finally, putative agents emerging from these secondary tests were clinically tested on humans. The report states that of the 34,500 compounds gathered as possible chemical agents, 2,077 entered the incapacitants' programme. Of these, 51 were selected for further testing and 32 were then recommended for clinical assessment in people. Clearly, any potential agent which went beyond that point would then be subject to much greater research and development effort in order to produce munitions and to ensure its safe utilisation by the military (see the lower half of Figure 8.1).

The Army study also attempted to estimate the total costs of this 'basic research and exploratory development of incapacitating agents' over the 25-year period. Despite considerable difficulties in collating data, it came to the conclusion that the programme had cost some $110 million. While that may not have been an enormous amount of money relative to the overall defence budget, it certainly allowed a great deal of research and development work to be carried out. For example, the investigators were able to find records of 25 contracts with 12 different contractors involving work on incapacitants, most of which appear to have involved the testing of agents on human subjects. The testing seems to have covered a wide range of chemicals and different types of subject. In 1961 North American Aviation received $51,840 to test pilot performance decrement after ingestion of BZ. In 1964 the University of Pennsylvania received $325,840 and tested inmates of Holmesburg Prison with 16 different chemicals such as **scopolamine** and 'various experimental **glycolate agents**'. Most bizarre was an experiment carried out under contract by Tulane University which 'involved giving **LSD** and **mescaline** to mental patients who previously had wire electrodes implanted in their brains'. It appears that the electrodes had been implanted before the military experiments as part of a medical research programme financed from other sources.

The atmosphere of excitement at that time about what might possibly be achieved in medical research is conveyed by a 1967 textbook in a sober account which began a chapter on psychotropic drugs thus:[11]

> During the past 15 years or so, the problem of the biochemical aetiology of mental illness and the mode of action of drugs which act on the central nervous system has become the object of intensive study . . .

It suggested that several factors caused this intense period of work and that:

> . . . Foremost amongst these events was the discovery of whole new ranges of compounds which had a powerful effect on emotional and behavioural responses . . .

These findings, combined with a growing understanding of neural transmission in the brain, led to the search for new or modified drugs with more precise properties.

Figure 8.1 Elements of the Incapacitants Search Programme*

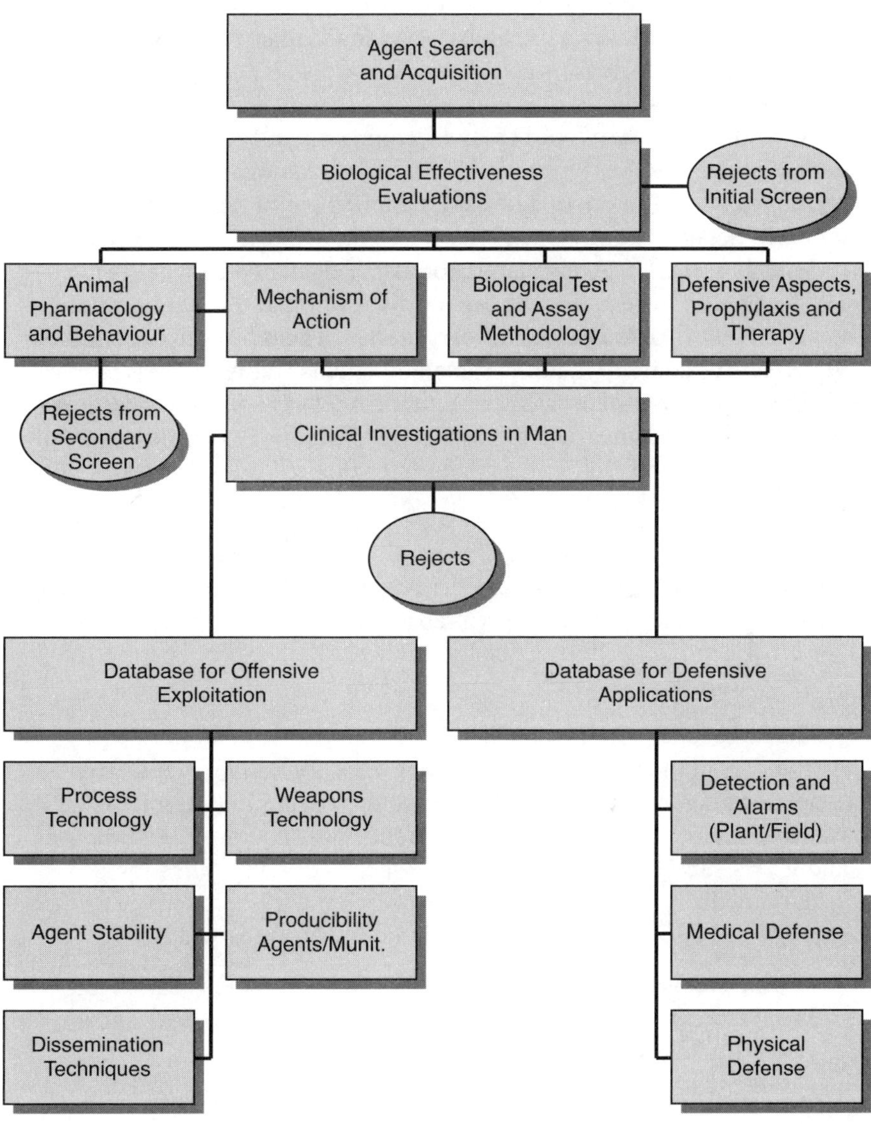

*From reference 10

In a sense then, the military work on incapacitants was not out of line with the medical research of the time, although it had a different purpose. However, as this medical text emphasised in its conclusion, working out what the effects of a drug are on the central nervous system is a very difficult task. As we have seen in Chapter 5, in spite of such difficulties the US Army was able to standardise BZ as an incapacitant agent – at least as an interim measure.

According to a follow-up study of volunteers in 1980, work on LSD took place between 1955 and 1967.[12] The hallucinogenic effects of this substance are well known and were sketched out in the previous chapter. Very detailed follow-up studies on volunteers who were exposed to chemicals at the US Army Laboratories at Edgewood Arsenal between 1958 and 1975 were carried out by the National Research Council Committee on Toxicology in the early 1980s. The reports give details of the effects of some of the other potential agents studied.[13]

Phencyclidine (sternyl, SN or SNA, EA 2148) was considered as a potential agent before the adoption of BZ. Volume Two of the report of the Committee on Toxicology sets out some of the features of this potential agent:

> SNA, an arylcyclohexylamine, belongs to a class of drugs known as dissociative anesthetics. Under the proper conditions of dosage and administration, it can act as a central nervous system (CNS) stimulant, a CNS depressant, a hallucinogen, an analgesic, or combinations of these. Related arylcyclohexylamines share some of SNA's pharmacological actions . . .

As we saw in the last chapter, the drug was originally developed as an anaesthetic agent for humans, but this use became impossible when it was discovered that the drug could produce post-operative disturbances and it was subsequently only used as a means of immobilisation in veterinary medicine. With the features it possessed, it is not surprising that variable effects were observed in volunteers who received it:

> . . . Excitation and arousal occurred in some and sedation, depression, and withdrawal in others at the same dose range – a well-known variability in response to SNA . . .

More complex disruption of the volunteers' minds also occurred:

> . . . feelings of estrangement and loneliness occurred in some, and an alcoholic-like euphoria occurred in others. Thought was generally disorganised. . . . Some subjects appeared catatonic and reported dream-like experiences at the highest doses . . .

Even worse:

> A schizophreniform psychosis can follow administration of single or multiple, usually high, doses of SNA. Allen and Young reported on nine patients seen over a 13-mo [month] period at an Army hospital with SNA psychosis within a week of multiple exposures. Despite anti-psychotic treatment, three had not recovered from the residue of their psychosis 30, 60, and 90d [days] later.

As the report points out, it was known that sternyl given to recognised schizophrenics could cause long-lasting psychotic symptoms. Still, the studies done at Edgewood were usually similar to tests carried out in civilian settings. Only the tests involving inhalation of sternyl – as would have been required in an actual incapacitant agent – were peculiar to the military. Sternyl was discarded as a potential agent in December 1959, largely because the minimum necessary dose was too large for field use.[14]

It will be recalled from Chapter 5 that from the early 1950s the Chemical Corps was interested in **tetrahydrocannabinols** – compounds related to the active chemical of the cannabis plant. Studies at Edgewood began in 1958 and lasted for a decade. It was found that **dimethylheptylpyran (DMHP)**, coded EA 1476, was much more potent than the natural product. Most work at Edgewood was done using an acetate form of DMHP, coded EA 2233. Between 1963 and 1966 about 100 volunteers were given doses of this substance. The Committee on Toxicology's report summarised as follows:

> . . . DMHP and some of its acetate isomers produced various degrees of physical incapacitation due largely to the moderate to marked and prolonged orthostatic hypotension [lowering of blood pressure]. . . . Mental effects of DMHP were much less severe than those of . . . cannabis at doses that produced similar degrees of orthostatic hypotension . . .

These effects on the cardiovascular system made many of the volunteers feel faint on standing. Whilst careful observations were made of the dose-to-effects relationship for oral and injected routes of administration, as was found in civilian studies, individuals could again exhibit quite variable symptoms for similar amounts of drug received.

Numerous reports from the early decades of the 50 years of research have subsequently become available to the general public. In 1971 the Human Engineering Laboratories at Aberdeen Proving Ground, Maryland, produced an annotated bibliography[15] of over 1,000 references concerned with 'Behavioural Modification and Changes in Central Nervous System Biochemistry'. In 1967 a third supplement to a contractors' report on 'New Incapacitating Agents' detailed the 'Preclinical Pharmacology and Toxicology of Candidate Agent 226,169',[16] and in 1968 a two-volume summary gave an indexed account[17] of the chemistry of 'New Incapacitating Agents'. In a seeming repetition of advances in military knowledge being passed on to police forces after the First World War,[18] the US Army proposed a series of disabling chemicals for police use in 1968.[19] Extracts from this listing (Table 8.1) give some idea of the knowledge available to the US military at that time. Extensive lists of CIA holdings of incapacitants,[20] and of the chemicals (including incapacitants such as LSD, BZ, Valium, chlorpromazine and fentanyl) tested by the Army and its contractors prior to 1975,[21] were also given at Senate hearings at that time.

The later years

From 1973 onwards it becomes an easier task to follow the outlines of the incapacitants development programme because, apart from a short break in the late 1980s when they were inadvertently removed by defence reorganisation legislation,[22] detailed semi-annual[23] and then annual reports on Chemical Warfare and Chemical/Biological Research Program Obligations[24] have been provided to Congress by the Department of Defense (DoD). Congress also receives descriptive summaries of the various elements of the programme as part of the annual DoD budget justification. In an August 1990 report[25] the General Accounting Office (GAO) concluded that the obligations report and the more detailed descriptions of the programme elements, which will be discussed later, were generally accurate and consistent with each other. However, it considered that the Department's reports did not

> . . . describe intermediate or overall program goals and objectives or set the accomplishments in the perspective of a broader purpose . . .

The DOD countered that it thought the information supplied was adequate for oversight, and if further information was supplied the report would have to be classified since vulnerabilities would be revealed. The GAO did not accept this view, but the reports available to us nevertheless lack the overview the GAO thought necessary for those without really specialised knowledge to understand properly the progress of the programmes reported on.

Nevertheless, the essentials of the process being reported may be followed by considering Figure 8.2. This diagrammatic representation of the programme flow is taken from the *Fiscal Year 1980 Arms Control Impact Statements* supplied to Congress.[26] As can be seen down the left

Table 8.1   Some of the disabling chemicals proposed in 1968 for use by police forces*

Respiratory irritants

    Capsaicin

Anaesthetics

    Sternyl and congeners

Analgesics

CS4640 EA3382	[Morphine given in higher doses than required for analgesia produces paralysis. Examples given had a poor safety margin in humans.]

Tranquillisers

Prolixin	[A phenothiazine; drug called fluphenazine.]
302,089	[A butyrophenone; the 3-methyl homolog of spiroperidol. Causes sedation then when sedation declines muscle spasms of neck, tongue and face. Said to inhibit desire to carry out tasks however strongly motivated a person may be.]

Glycolates

	[Cause physical weakness and confusion. Many had been synthesised and tested.]
EA3834 302,668	[Several-minute onset times and several-hours duration. EA 3834 two to three times more active. 302,668 also produces vomiting at lower doses than other glycolates.]
EA3167	[Several-hour onset and several-week duration.]
BZ	[Effects between those of EA3834 and EA3167.]
EA3580	[*ditto*]

Vomiting agents

Apomorphine	[Vomiting, then extreme nausea and weakness for two hours. Effective as an aerosol.]
228,926	[Structurally similar to apomorphine but more potent and therefore more useful for a dart payload.]

---

* From reference 19

side of the Figure, the process begins with research and then proceeds to exploratory development (XD). If this makes sufficient progress, a period of advanced development (AD) or even engineering development (ED) may follow. The various stages of the different parts of the programme are then given numerical designations. In 1980, for example, if advanced development of riot control agents (RCA) had been under way, it would have been in the programme element designated 63627A, as shown on the right side of the Figure.

Figure 8.2   CW RDT&E Program Flow*

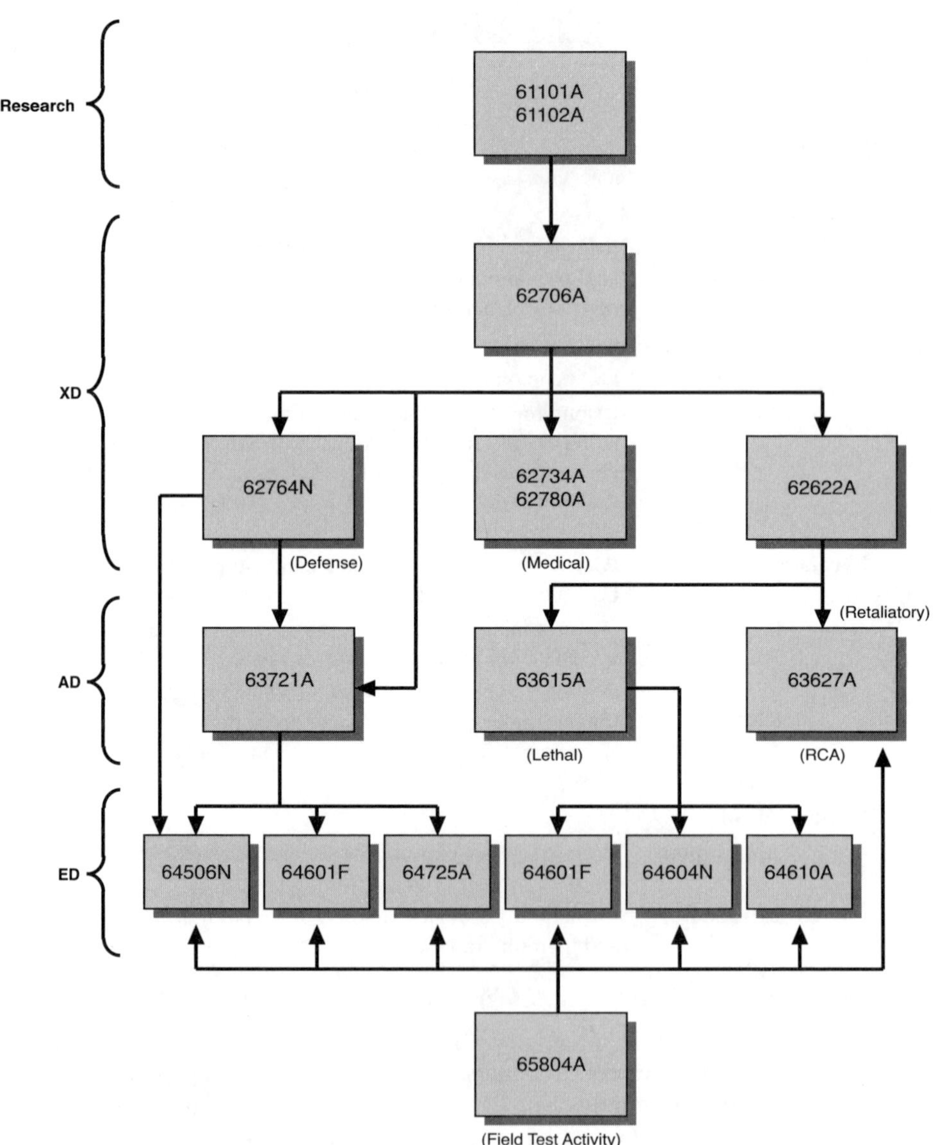

*From reference 26

One way to get an overall picture of the US Army's efforts to develop incapacitating agents and munitions since 1973 is therefore to determine when more than exploratory development took place. This information is set out in Table 8.2. The data are taken from Perry Robinson's chronological summary[9] and the available Congressional obligations reports (for example[23,24]). It would appear that we have two periods when incapacitants were moving towards deployment: the early to mid-1970s and the early to mid-1990s. There is, of course, some indeterminacy here because of the missing data for FYs 1987 and 1988. There may also be classified projects in this area, but for our purposes we may assume that there were just the two periods of movement towards deployment separated by almost a decade and a half of exploring possibilities. Though our main concern is with recent happenings, it is necessary to fill in something of the background from the earlier periods.

Table 8.3 sets out some extracts from the Congressional obligations reports for FYs 1973 to 1977. It will be noted that a number of different classes of chemical were still being investigated: analogues of **thebaine** and **oripavine**; **phenothiazines**; chemicals related to **morphine**; and **glycolates**. Moreover, there was interest in both respiratory and percutaneous routes of entry for the agents into the human body. It will be noted, from the extracts of the 'Basic Research in Life Sciences' programme for 1975, that funds were also being spent on related research outwith the incapacitants programme itself. Here we find investigations of brain structures and chemical pathways, and investigations of the effects of chemicals on aggression and memory.

From our viewpoint, of particular interest are the Army's approval of a requirement for a tactical, air-delivered, incapacitating munition system (TADICAMS) in 1973, and the efforts to produce both a 155mm XM-723 Incap. projectile and an XM-96 rocket warhead. As might be expected, there were also studies of the utility of a sub-munition (SUU-30B/B) for an incapacitant agent. By the end of this period, however, there were clearly problems with munitions, and tighter restrictions on testing in human subjects. It should also be recalled that by the end of this period the American involvement in Vietnam was over.

Table 8.4 provides similar extracts from the obligations reports for the period between FYs 1978 and 1986. During this time there was clearly a deepening understanding of the mechanism of action of potential incapacitants, and of how they might be weaponised and used.[27,28] The final report of a major study in 1984 of a 155mm munition loaded with sub-munitions stated, for example, that:[28]

Table 8.2   The progress of the incapacitants programme*

Fiscal Year	Exploratory development	Advanced development	Engineering development
1973	x	x	
1974	x	x	
1975	x		
1976	x		
1977	x	x	
1978	x		
1979	x		
1980	x		
1981	x		
1982	x		
1983	x		
1984	(x)[1]		
1985	x		
1986	x		
1987	n.a.[2]	n.a.	n.a.
1988	n.a.	n.a.	n.a.
1989	x		
1990	x		
1991	x		
1992	x		
1993	x	x[3]	
1994	n.a.	n.a.	

1. Not direct from Congressional Report (DMS Worldwide Market Study and Forecast [1986]).
2. Not available.
3. From PEDS.

* From reference 9

Drop tests were performed to experimentally test the capability of the individual pyrotechnic and liquid filled submunitions to withstand ground impact after expulsion from the projectile in flight.

Full-scale firing tests were performed at Yuma and Dugway Proving Grounds to verify the launch and flight stability of the round and the structural integrity of the payload.

Also of note are the interest in fentanyl[29] and the 'evaluation of potent *analgesics* and volatile *anesthetics*' [my emphases] at the end of the period (Table 8.4, 1985). More information is available from an

Table 8.3 Aspects of the search for incapacitants FYs 1973–77

Fiscal Year		Extracts[1]
1973	*	Synthesis of analogues and homologues of thebaine and oripavine.
	*	Preparation of additional quantities of phenothiazines.
	*	Studies continued on wide area coverage using a SUU-30B/B submunition for EA 3834.
	*	Requirement for a tactical air-delivered incapacitating munition system (TADICAMS) was approved by the Department of the Army.
	*	Significant progress made in development and design of 155 mm XM-723 Incap. projectile.
1974	*	Investigation of synthesis of incapacitating materials related to morphine.
	*	Several new phenothiazines prepared in sufficient quantities to begin biological evaluation.
	*	Some of the phenothiazines successfully thermally disseminated.
	*	XM-723 expected to transition to engineering development.
1975[2]	*	Experiments indicate that lesions of the posterior midbrain central grey area potentiate an antinociceptive action of morphine . . . it should be possible to identify some of the anatomical structures involved.
	*	Various neural pathways in the brain are known to be cholinergic in nature. It was of interest to learn whether some neural pathways which mediate pain sensation are also of a cholinergic nature. Laboratory tests to date indicate that this system is not cholinergic.
	*	Preliminary experiments indicate that Valium has potent anti-aggressive properties. Experiments are under way to evaluate the effects of other benzodiazepines and benzo-diazepine-anticholinergic mixtures on aggressive behaviour.
	*	The primary influence of scopolamine on memory appears to be in the impairment of acquiring new material.
1976	*	Available data on pyrotechnic incapacitating munitions reviewed with no solutions available to solve the safety problem of agent release during an accident.
	*	The 2-compartment thermal generation principal is the most promising dissemination method available.

Table 8.3   *(Cont.)*

1977	*	Because of the restrictions in testing potential incapacitating agents in man, major emphasis was placed on a review of incapacitants previously tested in volunteers.[3]
	*	This review was to select an improved chemical incapacitating agent that causes physical incapacitation through both respiratory and percutaneous routes of entry into body systems, and to develop specific methods for its dissemination so that incapacitating agent munitions can be developed to meet military requirements.[3]
	*	One approach to an improved incapacitant is the mixture of a volatile irritant with a glycolate compound.[3]
	*	A fill and close facility established for the 66 mm XM-96 rocket warhead.

1. Most extracts are taken direct from the obligation reports given to Congress on the 'Incapacitating Chemicals Program', but are sometimes truncated.
2. From the 'Basic Research in Life Sciences' section.
3. From reference 9.

Table 8.4   Aspects of the search for incapacitants FYs 1978–86

Fiscal year[1]		Extracts
1978	*	Studies conducted on two new series of compounds, one new compound was especially potent, but its safety ratio was too low.
	*	Two successful contract efforts on incapacitating pyrotechnic munition technology involved conceptual studies on safety aspects and munition redesign.
1979	*	Current emphasis of this programme has been on developing physically incapacitating agents that would not only be effective by inhalation, but would also penetrate clothing and be effective through the skin.
1980	*	Compounds structurally similar to the glycolates were evaluated and to date two appear to have pharmacological activity to merit further interest.
1981	*	A theoretical study of ways to relate chemical structure, pharmacological activity, and physical properties within a given class of compounds was completed.

Table 8.4   *(Cont.)*

1982	*	A structure-activity relationship study was initiated on a new class of compounds with focus on increasing both the safety ratio and the incapacitating effects.
	*	Two studies were initiated to improve the design of pyrotechnic projectiles for delivering incapacitating agents.
1983	*	A state-of-the-art review of potential incapacitating agents was completed and new classes of agents were selected for further study.
1984[2]	*	Studies carried out to explore the structure-activity relationships of carfentanil analogs and a series of bis-quaternary compounds.
	*	Fentanyl and its analogs identified through molecular modelling for future study.
	*	Redesigned the incapacitating 155 mm projectile and successfully fired it with non-toxic simulant.
1985	*	Objectives include synthesis and evaluation of potent analgesics and volatile anesthetics.
1986	*	Continued the search for new incapacitating agents utilising academic and industrial resources.

1. Extracts prepared as in Table 8.3.
2. *DMS WorldWide Market Study and Forecast* (1986)

Table 8.5   Incapacitant agent characteristics*

Agent	Human toxicity estimates
BZ	(ICt 50)[1] 112 mg-min/m^3
EA3834 (Glycolate)	(ICt 50) 73 mg-min/m^3
Phencyclidine	(ICt 50) 1,000 mg-min/m^3
EA5202 (Phenothiazine)	(MED 50)[2] 0.005 mg/kg
Morphine-type analgesics	
i) CS4640 (to knock down monkeys)	25 mg-min/m^3
ii) EA4929	(MED 50) 0.005 mg/kg
iii) EA5544	(MED 50) 0.0002 mg/kg

1. ICt 50 is the median airborne dose at which 50 per cent of those subject to attack are incapacitated.
2. MED 50 is the median injected dose at which 50 per cent of those injected are incapacitated.

* From reference 9

Edgewood Arsenal briefing in 1981,[30] on 'Characteristics of Chemical Agents', which listed toxicities of seven incapacitant agents (Table 8.5).

Understanding what has happened recently to the search by the US Army for effective incapacitants is made difficult because in FY 1990 it terminated the Incapacitating Chemical Program and initiated a Riot Control Program. Since it appears that the DoD does not regard riot control agents as chemical weapons, the FY 1991 obligations report contains no section on incapacitating chemicals.[9] To follow the programme we must therefore have resort to the more detailed Program Element Descriptive Summaries (PEDS) to Congress provided in the supporting data on Research Development Test and Evaluation documents during the 1990s.[31] Even in these documents, however, the progress of the programme is not easy to follow since it is switched from exploratory development project A554 to A552 in FY 1992.[9] We can, nevertheless, gain an insight into what was happening to the incapacitants' programme from the US Army's January 1992 *NBC Modernisation Plan*.[32]

This document explains that riot control devices come within the Flame/Incendiary and Non-Lethal (FINL) functional area in which innovative applications of technology are used to meet war-fighting challenges. Applications of interest are said to include those capable of.

> . . . destroying or disabling enemy armor and other vehicles; soldiers inside fortifications, bunkers, and foxholes with overhead cover; fighting systems inside fortifications, bunkers, and foxholes with overhead cover; soldiers inside buildings and structures in urban combat situations . . .

Flame/incendiary weapons are used to destroy enemy soldiers and *matériel* and to lower morale; anti-*matériel* weapons give a capability to destroy large area targets deep in enemy territory; and, according to this document, riot control devices are 'systems that produce performance degradation effects on humans'. It is suggested that in the mid-term (FYs 1998–2002) a non-lethal riot control agent, and hand-held and shoulder-fired delivery of such an agent, will support AirLand Operations. The envisaged modernisation strategy for FINL weapons is set out in Figure 8.3. The document gives the following background information on the development of an Advanced Riot Control Agent Device (ARCAD):

> A class of compounds has been selected and viable analogs are under evaluation for acceptability in meeting initial generic requirements. *The Acquisition Strategy (AS) was approved in February 1991. The Acquisition Plan was approved in May 1991* . . . [my emphasis]

On this evidence, there can be little doubt that a new incapacitant agent is due for deployment. The complex details of the progress of the 'Incapacitating Chemical Program', as described in the official reports since 1989, are to be found in Perry Robinson's chronology.[9]

For present purposes it is not necessary to follow these complexities. We can concentrate on those sections of the relevant reports to Congress which record what was accomplished in the search for an effective incapacitant during the 1990s. Table 8.6 therefore sets out some extracts from the obligations reports to Congress for FYs 1989 and 1990. It will be seen that in FY 1989 an opioid appears to have been selected, and studies were made of how to increase its safety by means of admixture with an adjunct compound. A decontaminant was also developed. In FY 1990 the 'Incapacitating Chemical' programme was terminated and the 'Riot Control' programme began. More detailed studies, for example of dissemination and inhalation, were then carried out.

Table 8.6   Aspects of the search for incapacitants FYs 1989–90*

Fiscal year	Extracts
1989 (Incapacitating Chemical Program: Exploratory Development)	
Project A554	
*	Awarded a contract for estimating the cost of production quantities of the candidate incapacitating chemicals.
*	Completed intermediate toxicity studies of the candidate opioid with an additional animal species.
*	Initiated studies with a opioid-adjunct mixture for enhanced safety.
*	Developed a decontaminant for the candidate incapacitant.
1990 (Incapacitating Chemical Program: Exploratory Development)	
Project A554	
*	Terminated the Incapacitating Chemical Program and initiated Riot Control Program.
*	Initiated inhalation studies for the candidate riot control material.
*	Initiated experimentation to evaluate dissemination techniques of candidate material.
*	Initiated studies to improve methods of synthesis and produce quantities required for the program.
*	Initiated compatibility and stability tests for each of the candidate riot control components.

* From Obligations Reports to Congress

Figure 8.3   Modernisation Strategy Flow
Flame/Incendiary/Non-Lethal Weapon Systems*

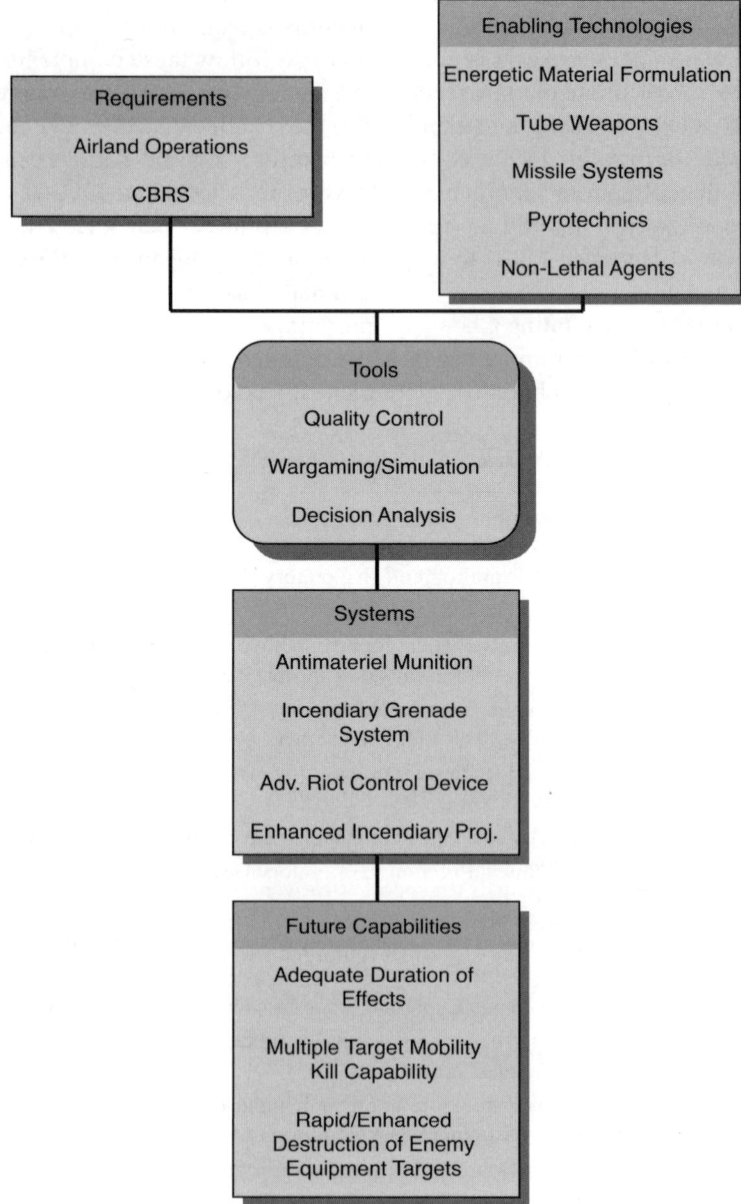

*From reference 32

We can take up the story again from the relevant historical reporting sections of the PEDS for FYs 1991–95. In FY 1989 (Table 8.7) it will be observed that work was also underway on adrenergic compounds. By FY 1990 there are clear signs of movement towards production involving market surveys for materials. By FY 1993 (report dated February 1994), it was recorded that work had progressed from exploratory development to advanced development. Under Program Element 0603627A and Project DE78, which provided 'Advanced Development of non-lethal riot control devices', we can see that the FY 1993 accomplishments included the completion of a preliminary manufacturing plan for the Advanced Riot Control Agent Device (ARCAD). Plans for FY 1994 then moved on to prototype design and hardware fabrication.

Table 8.7   Accomplishments FYs 1989–94*

FY 1989 (Project 554)

* Synthesised small quantities of candidate incapacitant agent and intermediates for physical-chemical and toxicological testing.

* Completed preliminary structure activity correlation model for adrenergic compounds.

* Completed physical property measurements for candidate incapacitant compounds.

FY 1990 (Project 554)

* Completed preliminary modelling study for dissemination of an Advanced Riot Control Agent (ARCA) in enclosed area.

* Completed market survey for prime ARCA materials.

* Evaluated candidate ARCA compounds by defined screening mechanisms and toxicological testing.

* Synthesis and determination of physical properties of selected ARCA compounds to support effectiveness analysis, dissemination studies and weapons system selection.

FY 1993 (Project DE78 – Advanced Development)

* Conducted Advanced Riot Control Agent Device (ARCAD) configuration analysis.

* Updated ARCAD Testing and Evaluation Master Plan (TEMP)

* Completed preliminary manufacturing plan for ARCAD.

FY 1994 (Project DE78 – Advanced Development)[1]

* Award ARCAD development contract.

* Initiate ARCAD prototype design and hardware fabrication.

---

* Data from PEDS
1. Planned

What is not clear from the published information is exactly what agent (or class of agent) the US Army has chosen for the device at this time. A look at the record of their interests over the last 50 years and a comparison with the developing knowledge of brain chemistry gained in civilian medical science suggest that their basic strategy has been to exploit potential agents which arose from other areas of research. There seems to have been a close parallel between civilian developments and military programmes. Given the interest in fentanyl and its derivatives in the US military during the 1980s (Table 8.4, 1984 entry), one might hazard a guess that the statements under Project A554 in 1989 (Table 8.6, first section), noting studies on 'the candidate opioid' and on an 'opioid-adjunct mixture for enhanced safety', might have some significance. But from our point of view the more interesting question is where this programme may be leading.

It is pertinent first to consider what other countries might think of the US ARCAD. An article in *Armada International* in 1990 considered the subject of 'Chemical Warfare Today' from a straightforward analytical/military position. The author noted that:[33]

> . . . BZ is supposed to remove a soldier's will to fight but during trials it has operated in erratic ways. Some soldiers do indeed become nervous wrecks while others cannot notice any difference. So BZ is, at best, an unknown factor, *but it is virtually certain that there are other similar agents, if not in service at least under development* . . . [author's emphasis]

In the popular perception, chemical weapons are important because they can cause massive casualties. To the military, however, this potential outcome can be prevented by the detection of agents and the donning of protective suits. The problems for the military are the resources consumed in dealing with the *possibility* of chemical attack, the troops which have to be allocated to detection and decontamination activities, the difficulties of working effectively in protective suits and so on.[33] Thus, from a military opponent's perspective, US development of an incapacitant capability might look much more threatening than the label 'non-lethal riot control agent' might imply. This is because of the resources that would have to be committed to counter the threat, which would be to the detriment of war-fighting capabilities.

Any military analyst would also have to consider possible attitudes among other forces to the requirement for a high safety ratio in an incapacitating agent. Many chemicals have been screened as potential incapacitants and a prime consideration in the US programme has been to find agents with a large gap between the effective and the lethal dose so that the agent could be used with little risk of permanent harm to those affected. There are probably many agents with excellent incapacitant properties which have been rejected because they lacked a satisfactory safety ratio. But another state might argue that greater risks were worth taking, in order, say, to incapacitate dangerous people in a target area, because innocent people outside the area would be less at risk through reduced agent concentration off target. Some very dangerous chemicals could then be fielded in the guise of riot control agents which, to military analysts, might have many of the drawbacks associated with lethal chemical weapons.

Perceptions of another country's research and development programmes can be equally problematic. Susan Wright and Stuart Ketcham attempted to analyse how the US Biological Defense Research Programme might be interpreted by others. One example they considered was the Medical Research and Development Command's interest in an opioid-like hibernation trigger substance. Specifically, they discussed a project funded by the command which:[34]

> . . . aims at the isolation and characterisation of the gene for a protein called Hibernation Induction Trigger or HIT, and the mass production of the protein using genetic engineering. HIT, a substance isolated from the blood of hibernating animals, was shown . . . in earlier studies to induce a hibernation-like state . . .

Given well-known official concern over the possible military misuse of bioregulators,[35] how might such military-financed research appear to other countries? Wright and Ketcham suggest that despite claims made for medical applications of HIT:

> It is difficult to see how mass production of HIT would contribute to biological defence. Offensive application, on the other hand, seems more likely if delivery of HIT as an active agent is achieved . . .

On balance then, while such research might be intended as defensive, it has the potential to be misunderstood and misused. Bearing such concerns in mind, we must now attempt a glimpse into the future.

Incapacitant Futures

The Chemical and Biological Defense Command (CBDCOM) of the US Army was established on 1 October 1993. Its headquarters are in the

Edgewood area of Aberdeen Proving Ground, Maryland and its mission is to 'handle research, development, acquisition and remediation issues associated with chemical and biological defence'.[36] The Edgewood Research, Development and Engineering Center (RDEC) is the largest CBDCOM element and its mission is to:[37]

> . . . conduct and manage basic and applied research, development, engineering and sustainment for chemical and biological defence systems and equipment.

Aberdeen Proving Ground contains many more US Army elements. One of these is the US Army Medical Research Institute of Chemical Defense (MRICD). This is a subordinate element of the US Army Medical Research and Materiel Command of the Office of the US Army Surgeon General. Scientists at MRICD:[37]

> . . . conduct research to characterise the effects of various chemical warfare agents and selected biological neurotoxins. They seek to define what biological systems are affected by the agent or toxin, and what short- and mid-term consequences of exposure are.

It seems reasonable to assume that any openly-available information on the work of Edgewood RDEC and MRICD could at least provide some indications of what agent types there might be in the future.

RDEC annual conferences

It is possible to gain an idea of current publicly accessible work on incapacitants from the annual conferences held by those involved in such work at Edgewood RDEC. The work reported at the conferences is diverse and we are only interested in part of it here. Some relevant titles and abstracts for the conferences from 1989 to 1994[38] have been rearranged, according to what seem to be two important research areas, in Tables 8.8 and 8.9. Whatever the nature of the agent currently in the weaponisation process, there obviously remains an interest in **opioids** and particularly **fentanyl** and its derivatives (Table 8.8). The investigators involved within Edgewood RDEC and contractor groups are using advanced mechanistic chemistry and achieving some clear results (see the emphasised sections of the extracts).

Most molecules in living organisms are based on the element carbon and carbon is often found in ring systems such as the well-known benzene ring, with its six carbon atoms. A carbon ring system in which one of the six carbon atoms is replaced by a different atom is termed a heterocycle. **Piperidine** is one such example.[39] Its six-membered ring has

Table 8.8   Edgewood RDEC studies of fentanyl and associated chemicals*

Year	Abstract No.	Title	Extract
1991	36	Synthesis and bio-activity of 2-(α-hydroxy-p-alkoxybenzyl) and 2-alkoxyarylamino analogs of etonitazene	'Etonitazene . . . is an opioid analgesic in a class of compounds whose potency appears to be sensitive to minor structural changes . . .'
1992	133	Approaches toward synthesis of 4-substituted-3-piperidones	'Research efforts over the last thirty years have resulted in the remarkable discovery that certain 4-anilidopiperidines have analgesic potencies thousands of times that of morphine. The *cis-3-methoxy derivative of fentanyl has recently been shown to be almost 30 times more potent than fentanyl . . .'*
1993	31	Synthesis and bioactivity of 2-fluoroanilide congeners of carfentanil, sufentanil and alfentanil	'. . . Preliminary pharmacological evaluation in mice has shown the fluoro analog of carfentanil to be about eight times more effective than the unsubstituted structure for a variety of signs. *The LD [Lethal Dose] 50 remained about the same, hence the safety ratio was also improved by a factor of about eight.*'
1994	37	Molecular mechanics calculations applied to the solution of synthetic problems	'. . . molecules of interest today are best handled in terms of more sophisticated models that are accessible only through molecular mechanics and other more rigorous calculations. *The results of recent calculations that correctly predict the outcome of reactions within the piperidine family of compounds will be presented . . .*'

* From reference 38

one carbon atom replaced with a nitrogen atom. **Morphine** is a complex molecule which[40]

> . . . can be regarded as a derivative of phenanthrene, *piperidine*, dibenzofuran, isoquinoline, benzodihydrofuran, furan, and other ring systems . . . [author's emphasis]

Morphine has obviously been the subject of many investigations to discover how its structure determines its specific properties and how these may be modified. **Codeine**, also found in the opium poppy, is the natural precursor of morphine. Its structure is therefore very similar to morphine, yet[40]

> . . . This relatively minor structural alteration results in a marked lowering of analgetic potency as well as dependence liability . . .

This fact allows the widespread medicinal use of codeine. A large series of piperidine derivatives was produced from the late 1930s including pethidine, which became widely accepted as a substitute for morphine in treating moderate pain. The piperidine ring, in fact, is said to be 'the most frequently encountered heterocycle in pharmaceutical agents'.[41] **Fentanyl** itself may be classified as a 4-anilopiperidine and, given its widespread use as an anaesthetic,[42] could be of interest to the Army as a potential incapacitant.

The second set of abstracts seems to indicate where the action may be in the future (Table 8.9). As the 1994 quotation from abstract number 33 noted:

> . . . More selective α_2-*adrenergic compounds* with potent sedative activity have been considered to be ideal next generation anesthetic agents which can be developed and used in the Less-Than-Lethal Technology Program . . . [author's emphasis]

This statement might possibly be intended for public consumption only, but even if that is the case, much of Edgewood's expertise is obviously being deployed in this area of adrenergic receptor research and it would presumably only be deployed in what was considered a generally useful direction. Even if, then, the quotation overstates the case, we can reasonably investigate the effort devoted to α_2-adrenergic receptors as an illustration of what might come about in the future.

The investigators involved are from both RDEC and outside contractors. The team's objectives are quite clear from the parts of the abstracts highlighted in Table 8.9. It appears to be the standard story of a discovery by others of a new class of compound with interesting

Table 8.9 Edgewood RDEC studies of α_2-adrenergic agonists *

Year	Abstract No.	Title	Extract
1989	41	Medetomidine analogs as α_2-adrenergic agonist	'. . . At the present time its [medetomidine] sedative and hypotensive effects seem to be manifest in the same dose range. *We have initiated a program to see if it is possible to separate these activities with analogs of medetomidine . . .*'
1990	140	Synthesis of new medetomidine analogs as α_2-adrenergic agonists	'Medetomidine is reported to be a highly selective and potent α_2-adrenergic stimulant. . . . *Our long range plan is to achieve pharmacological selectivity by chemical modification of medetomidine.*'
1991	26	α_2-adrenergic activity of a new series of analogs of 4-[1-(1-naphthalenyl) ethyl]-1H-imidazole	'*Due to the potent α_2-adrenergic activity of 4-[1-(1-. . . an analog of medetomidine, we have initiated a structure activity relationship study of this compound.*'
1992	134	Synthesis of naphthalene analogs of medetomidine as α_2-adrenergic agonists	One of the 4-substituted imidazoles, medetomidine, is reported to be highly potent on α_2-adrenergic receptors. *Both sedative and hypotensive effects are noted with medetomidine.* We have recently shown that a naphthalene analog of medetomidine is also a very potent and selective α_2-adrenergic stimulant . . .'
1993	26	Synthesis of conformational restrained analogs of medetomidine as α_2-adrenergic agonist	'. . . *A series of 4-substituted imidazoles attached through a methylene bridge to a 1-naphthalene system has been prepared.* The type of substitution on the methylene bridge seems to be critical for obtaining α_2-agonist activity . . .'

Table 8.9   *(Cont.)*

Year	Abstract No.	Title	Extract
1994	32	Synthesis and biological activity of a series of conformationally restricted analogs of 4-substituted imidazoles as α_2-adrenergic agonists	'. . . *As part of a search for a new agent* of the medetomidine series which shows potent anesthetic, analgesic and sedative activity with a lack of adverse cardiovascular side effects, *we discovered that a naphthalene analog of medetomidine has greater potency as well as higher selectivity for α_2-adrenergic receptors . . .*'
	33	Synthesis and α_2-adrenergic activity of quinoline and quinoxaline analogs of medetomidine	*'Centrally acting α_2-adrenergic compounds show antihypertensive actions with sedative properties. More selective α_2-adrenergic compounds with potent sedative activity have been considered to be ideal next generation anesthetic agents which can be developed and used in the Less-Than-Lethal Technology Program. Unlike opioids, these* compounds are devoid of the usual liabilities associated with respiratory depression, physical dependence and environmental concern after dissemination . . .'

* From reference 38

actions on the nervous system and then a systematic attempt to generate related chemicals which might have more selective properties. In this case the new compounds were found to have strong effects on α_2-adrenergic receptors and the key chemical was **medetomidine**.[43] As the 1990 abstract number 140 noted, 'our long-range plan is to achieve pharmacological selectivity by chemical modification of medetomidine' while the 1994 abstract number 32 stated:

> . . . As part of a search for a new agent. . . . we discovered that a naph-
> thalene analog of medetomidine has greater potency as well as higher
> selectivity for α_2-adrenergic receptors . . .

Another aspect of this research is important to grasp. Unlike the researchers of the 1950s, the team is now able to build on many years of research, in the whole scientific community, on the structure of receptors and of chemicals that may affect them. Backing up this research, Edgewood also has access to vastly improved computing capabilities that could begin to give predictive estimates of structural relationships to greatly speed the pace of research (see the 1994 abstract number 37 in Table 8.8).

Given that brief overview of the work of Edgewood RDEC on α_2 adrenoceptors, how do we assess its significance?

Inducing sedation

Medetomidine was originally introduced as a sedative and analgesic drug for veterinary use by the Farmos Group in Finland.[43] The same group demonstrated that the activity resides in the (+) optical isomer – *dex*medetomidine.[44] In humans dexmedetomidine causes a transient increase in blood pressure (via a peripheral mechanism) and then a centrally-induced lowering of blood pressure and heart rate. It causes sedation at doses ranging from 0.25 – 2 µg/kg delivered intravenously. At the upper end of this range some subjects could not be aroused for a brief period by vocal commands and sedation effects persisted for 190 minutes.[45] This effect, at 0.000002 gms per kg of body weight, would appear to be well within the limits suggested as appropriate for an incapacitant in Chapter 5. Yet as we would now expect, the effect of the same dose on different people may be quite variable.[46]

The possible future use of such chemicals in clinical anaesthesia was subject to a detailed review in the journal *Anesthesiology* in 1991.[47] The review noted that administration of α_2 adrenoceptor agonists produced effects on the neuroendocrine, renal and gastrointestinal systems in addition to the marked effects on the cardiovascular system and the central nervous system. Alpha-2 (α_2) adrenoceptors have been found quite widely distributed in the CNS, but:[48]

> The localisation of α_2-binding sites in the human brain provides an
> anatomical basis for the known pharmacological effects of clonidine
> and other centrally acting α_2-agonists . . .

This can be demonstrated easily for regions known to be involved in control of the cardiovascular system, but here we shall continue to concentrate on the control of sedation.

It has been found that the distribution of α_2 binding sites correlates to a considerable degree with the known location of norepinephrine in the brain, and it will be recalled from Chapter 7 that a large number of the norepinephrine-containing cell bodies are located in just two groups – the bilateral **locus coeruleus** (LC). Modern anatomical techniques have allowed much information to be gathered on the development of these norepinephrine cells of the locus coeruleus in a number of mammalian species, including humans, and on how they come to innervate very large areas of the central nervous system.[49] A review in the mid-1980s of the functional implications of the anatomy and physiology of this group of just several thousand neurons (out of billions in the brain), noted that:[50]

> . . . At the present time the NE-containing locus coeruleus (NE-LC) system is probably the most fruitful focus for functional speculation because more is known anatomically, physiologically, and clinically about this particular NE neuronal system than any other in the brain . . .

In short, this is the prime neurochemical system in the brain which we may hope to understand at present.

Many theories have been put forward for the functions of the NE-LC but the reviewers concentrated on the seven they considered to have most support – its role in: arousal and affective disorders; learning or memory; reinforcement; sleep; anxiety or nociception; cerebral blood flow and metabolism; and attention. They concluded, however, that the best working hypothesis was that:

> . . . enhanced activity traversing the elaborate efferent projections of the NE-LC system (as a result of enhanced external signals) biases the orientation of the brain and behavioral activities so as to deal preferentially with external events of phasically high priority . . .

On the other hand, 'diminished NE-LC activity would allow internally orientated, more instinctual brain and behavior programs to be expressed'. We shall return to this theory shortly.

From our standpoint, what is important is the demonstration that the action of dexmedetomidine on the locus coeruleus *alone* can induce sedation in rats. The experiments involved introducing a cannula into the brain so that the drug could be injected directly on to the LC neurons. A dose-dependent sedation was achieved when dexmedetomidine was injected onto just one of the bilateral cell groups. Injection of the same substance a little distance away from the LC neurons did not sedate the animals, and successful sedation could be prevented by injecting both dexmedetomidine and a specific α_2-adrenergic antagonist onto the LC. The authors of this study suggested a complex system in which:[51]

> . . . it is probable that the spontaneous discharge of noradrenergic pathways relaying from the LC are excitatory, perhaps by attenuating inhibitory interneurons . . .

The mechanism involved is therefore complex, but understandable:

> . . . Activation of the α_2 adrenoceptors suppresses the spontaneous firing rate of the LC. Thus, interruption of noradrenergic neuro-transmission by lesioning or by direct activation of the α_2 adrenoceptors . . . may disinhibit the inhibitory interneurons, such as the GABA-ergic pathways, to produce central nervous system depression.

Clearly, when dexmedetomidine is used globally, the LC neurons are not the only system affected, but as the authors concluded:

> . . . Now that we have localised a site for the α_2 adrenoceptor-mediated hypnotic response, we are able to deposit selective toxins and antibodies . . . to investigate the molecular components involved . . .

It would appear that the long-sought goal of a specific chemical to produce sedation at low dosage by a known mechanism draws near – particularly since the genes encoding human α_2-adrenoceptors have been identified and cloned so that their properties can be investigated in detail.[47] More recently, it has been shown that the *sedative* action of medetomidine mediated via the LC in rats can be differentiated from the *pain reduction* response which is produced elsewhere.[52] Similar results had previously been obtained in studies of the affect of clonidine on locus coeruleus neurons in the rat.[53] It should additionally be noted that opiates, acting through other receptors, have also been found to depress LC neuron activity. Indeed, opiates (eg fentanyl) and α_2-adrenergic agonists (eg clonidine) can be used synergistically to produce anaesthesia.[47]

With this history of successful research, it is hardly surprising that many increasingly sophisticated studies on the locus coeruleus system continue to be published. It is not necessary to discuss this work here but we need to note one important issue concerning the interest of armed forces in the subject. In one study the investigators recorded activity from single LC neurons in unanaesthetised monkeys and showed that there was a clear correlation with the behavioural state of the animal in a way they would have expected. For example:[54]

> . . . In alert waking, LC neurons displayed continuous, moderately irregular activity. In contrast, prolonged pauses in activity accompanied drowsiness. These pauses preceded eye closure and occurred 1-3s [seconds] before the onset of slow-wave EEG. At awakening, LC activation preceded by up to 3s desynchronised EEG and eye opening . . .

Other results showed LC neurons firing only in response to an infrequent visual signal the monkey needed in order to carry out the action required to get a juice drink! What is interesting, however, is that the study was sponsored by the US Air Force Office of Scientific Research. This and related Air Force-funded research, details of which were obtained from a computer database search of publicly available literature, are listed in Table 8.10.

It is possible to argue that there is nothing unacceptable in US Air Force funding of such work. The Air Force openly states that in its Neuroscience Programme it wishes to understand the neural mechanisms which support effectiveness in carrying out skilled tasks in demanding circumstances and that:[55]

> . . . Areas of emphasis are fundamental studies of the neurobiological mechanisms underlying neuronal responsiveness, learning and memory, fatigue, stress, attention and arousal.

It can also be argued that the work is available in the open literature, as we have seen. Moreover, knowledge of medetomidine is widespread among veterinarians around the world (see, for example, reference 56), and more advanced studies of the use of such substances, under US public health funding, are being shared with scientists from other countries (see, for example, reference 57). Yet it may be stretching the limits of credibility to argue that US work financed in this way will always be perceived in a benign way in other countries. In view of the oft-stated determination of the US military to retain a technological edge, and the belief of other major states that they have well-founded reasons for caution about Western intentions, there are general grounds for thinking that such countries may be somewhat suspicious.

So it is hardly far-fetched to imagine that other countries may follow the US search for incapacitants and perhaps, as we have surmised, with less concern for the gap between effective and lethal dosages. Given that:[58]

> . . . Every time that lethal chemicals have ever been used, it started with [non-lethal] tear gas . . .

there are surely some grounds here for concern. As will become

Table 8.10   USAF-funded research[1]

Year	Title	Grant Number[2]
1994	Locus coeruleus in monkey: Phasic and tonic changes are associated with altered vigilance	AFOSR-90-0147, F49620-93-1-0099
1994	Locus coeruleus neurons in monkey are selectively activated by attended cues in a vigilance task	AFOSR-90-0147, F49620-93-1-0099
1994	Phasic stimulation of the locus coeruleus: effects on activity in the lateral geniculate nucleus	AFOSR-90-0294
1992	Acute morphine induces oscillatory discharge of noradrenergic locus coeruleus neurons in the waking monkey	AFOSR-90-0147
1991	Afferent regulation of locus coeruleus neurons: Anatomy, physiology and pharmacology	AFOSR-90-0147
1991	Discharge of noradrenergic locus coeruleus neurons in behaving rats and monkeys suggest a role in vigilance	AFOSR-90-0147
1991	Single-unit and physiological analysis of brain norepinephrine in behaving animals	AFOSR-87-0301-A
1991	Projections from the periaqueductal gray to the rostromedial pericoerulear region and nucleus locus coeruleus: Anatomical and physiological studies	US Army DAMD 17-86-C-6005[3]
1990	Locus coeruleus activity in behaving animals	AFOSR-87-0301
1990	The role of brain serotonin	AFOSR-85-0034

1. Data from computer database searches.
2. Only AFOSR funding listed except for footnote below.
3. Department of the Army, not AFOSR.

apparent in Chapter 9, such concerns multiply when the issue of chemical arms control today comes under discussion.

At the Edgewood Research, Development and Engineering Center Annual Scientific Conference on Chemical and Biological Defense Research in November 1995 a paper was presented summarising 40 years of work on 'less-than-lethal' chemicals at Edgewood.[59] The abstract of the paper argued that these chemicals were among the most mature of the available non-lethal technologies and that:

> . . . Depending on the specific scenario, several classes of chemicals have potential use, to include: potent analgesics/anesthetics as rapid acting immobilisers; sedatives as immobilisers; and calmatives that leave the subject awake and mobile but without the will or ability to meet objectives . . .

These chemicals were judged to have applications in both military missions and law enforcement. Here, if any further evidence is required, is confirmation that the US is now in possession of what it considers to be effective, modern, non-lethal chemical agents.

.9.

ARMS CONTROL FOR THE 21ST CENTURY

Arms control cannot deal directly with the roots of many of the international social conflicts that increasingly require the attention of peacekeepers and peacemakers, but failures of arms control can certainly exacerbate, and render more lethal, degenerating social conflicts. Arms control may therefore be regarded as one of the set of preventive international security responses identified by the Australian Foreign Minister, Gareth Evans, as requiring greater emphasis in the future:[1]

> . . . To tackle the problem of intrastate conflict more constructively means rethinking the doctrinal foundations for international security responses; giving much greater emphasis than hitherto to preventive, as distinct from corrective, strategies; and giving much more serious and sustained attention to organisational reform, particularly within the UN.

This chapter will review some of the necessary rethinking of arms control which is currently being pursued.

During the second phase of the Cold War in the early 1980s many people despaired of the idea of arms control as a mechanism of restraint on militarisation. The fundamental East–West antagonism appeared to block any political progress, while the application of vast sums of money and outstanding talent to research and development ensured a continued globalisation of defence industrial capability and a new military-technological revolution. Since the late 1980s, as political conditions have changed and defence budgets have declined, arms control has certainly been a more successful enterprise and some arms control specialists have attempted to suggest how it might be reconceptualised to operate more effectively within a new world order. We shall begin by looking at some such efforts at reconceptualisation from within the 'traditional' arms control community. However, the globalisation of defence industrial capability continues apace, particularly as many new and important technologies have dual – civilian and military – applications. We must therefore also briefly review the current state of the military-technological revolution and its spread around the world.

While the Cold War period witnessed a divide between the concerns of international lawyers and arms control specialists from the strategic studies area, this was only an extreme manifestation of the rift between international lawyers and the whole of the international relations community of scholars whose dominant realist orientation just did not allow a significant role for international law. More recently, however, there appears to have been a fruitful *rapprochement* between international relations scholars and lawyers in the analysis of international regulatory regimes and the associated development of norms and implementation through organisations. The second part of this chapter will therefore be concerned with global prohibition regimes, and will raise the question of whether the Cold War notion of arms control should now be superseded by the older idea of genuine disarmament.

We then turn to the specific example of the Chemical Weapons Convention (CWC) which, it has been suggested, is the paradigm case of the new form of international legal regulation that we need for the control of worldwide militarisation. The question is how such an advanced regulatory format might be applied to the control of proliferation of non-lethal weapons? Here we also touch upon the complex question of how norms of behaviour required to reinforce global prohibition are formed and maintained. Finally, we consider the question of implementation within the United Nations system and ask whether it can develop an effective multilateral negotiation mechanism for generating the new regimes that we appear to require.

Reconceptualising Arms Control

Criticism of arms control during the second phase of the Cold War came from two distinct sources in the West. Some, on the political right, feared that arms control merely provided a means by which totalitarian states could more easily threaten the security of democratic states by cheating on agreements. Others, on the left, feared that arms control agreements merely legitimised an arms race in which both East and

West were involved. More surprisingly, given the arms control successes of the late 1980s and early 1990s – for example, the INF and the CFE – serious concerns about the usefulness of arms control continued. As Lawrence Freedman noted, in the special *Daedalus* volume on arms control in 1991:[2]

> . . . there is a particular poignancy here for arms control in that on the eve of its greatest triumph its relevance is widely questioned and the political events which made rapid progress possible may now be moving so fast as to whisk the triumph away from its grasp . . .

Freedman's was one of many analyses made by specialists of arms control asking, essentially, whether it still had a role to play. His point was that arms control thinking in the West had been dominated for 30 years by Schelling and Halperin's definition, which primarily addressed the issue of nuclear stability between the superpowers. Any listing of the arms control treaties of the period shows this quite clearly.[3] With the ending of the Cold War, arms control has become much more complex, and the standard theories no longer appear to fit the problem very well.

An interesting example of a theoretical reconceptualising of arms control was Ivo Daalder's contention in 1992[4] that we should consider its function within a continuum of political relations between states, in which there is war at one end and peace at the other. Clearly, there is a need rather for humanitarian laws of war at one end and no need of formal arms control treaties at the other. In between, however, various forms of arms control might operate. Whilst traditional forms of (competitive) arms control would exist where there were grave political differences and concerns were dominated by the security dilemma, a second (co-operative) form:

> . . . has as its objective the improvement of political relations among states by creating, through an intensive dialogue and habitual co-operation, a set of principles, norms and rules that govern the military dimension of interstate relations . . .

Thus where such a form of arms control might operate, a cycle of reassurance and reciprocal arms build-down could replace the arms build-up associated with the security dilemma.

At a practical level, Brad Roberts suggested in the same year that the United States would have three 'focal points' in its arms control policy for the decade ahead.[5] These would be: a continuation of the old East–West agenda; a series of global multilateral issues; and some important regional problems of arms control. The old East–West agenda, in

Roberts' opinion, requires: continued attention to ensure that existing agreements such as CFE and START are implemented; that, if possible, follow-on treaties to cut nuclear arsenals further are achieved; and that new agreements on European security are agreed if necessary. At the global level, he argued that the decline of the 'bipolar world order' would lead to increasing reliance being placed on multilateral agreements such as the Non-Proliferation Treaty (NPT) and the Biological Weapons Convention (BWC). Finally, he noted the difficulties of achieving arms control measures in conflict-prone regions such as Korea, India and Pakistan, and the Middle East. Moreover, he argued that while strong regional agreements could enhance global regimes, weak regional agreements could eventually erode global efforts as well as contribute to regional insecurity at times of crisis.

Such examples demonstrate that arms control thinking in the West during the Cold War largely addressed only a part of the range of functions for which the historical record shows arms control has been used.[6] The massive arms control agreements put in place at the end of the Cold War serve to remind us that the predominant paradigm during the Cold War had narrowed down to a consideration of arms control almost exclusively in regard to the maintenance of strategic nuclear stability between the superpowers. Other possible functions of arms control were peripheral. The agenda from 1985 onwards, in contrast, was centred on achieving a new arms balance at the end of the long confrontation. The question that naturally arises is, 'What is the arms control agenda appropriate to the start of the next century, when the present period of transition from the Cold War has, perhaps, run its course?'

In his thoughtful article, 'The Difficult Transformation from "Arms Control" into a "World Security System"', Bertrand[7] may not have answered that question, but he did detail some of the points that require careful consideration in order that an answer may be provided. He asked, for example, what the dominant features of the new system will be, what kinds of threat will be important, what means might best be used to address such threats (including the types and sizes of armed forces and the measures of arms control), and what processes and institutions might be most effective in reaching the necessary agreements within that system.

While this is not the place to attempt a full answer to the question, two features of the world system within which we will have to work over the first decades of the next century are surely quite clear. First, as argued in Chapter 1, whatever hopes there may initially have been for a new world order following the ending of the Cold War, we shall surely experience a world in which conflict and war remain endemic. In particular, the imbalances of wealth between rich and poor, and the limits to the resilience of the natural environment, are likely to result in extensive intra- and inter-state conflict in the South, and between North and South.[8] Secondly, the military expenditures of the Cold War period have resulted in such vastly increased military capability – in lethality and precision, for example – as amply to justify the terminology of a 'military-technological revolution' in combat possibilities between regular forces.[9] Any doubts on that point were dispelled by the US-led assault on Iraq in 1991.[10] But more important, probably, is the fact that the industrial capabilities required to sustain such new warfare are spreading around the world, in part because the defence industries of the North are in trouble as defence expenditures decline and are thus being forced to seek new markets and provide very advantageous terms of trade to arms buyers.[11] For these reasons the necessity of controlling the proliferation of advanced military capabilities has become central to the arms control agenda for most analysts in the West. Particular concern is directed at the proliferation of weapons of mass destruction and the means of their delivery, and the difficulty of encouraging participation by poorer states in control regimes through co-operation in civil technology while, at the same time, preventing military misuse of what are frequently dual-use industrial capabilities.[12]

It is important to note that discussions of the proper role of arms control, if any, in constructing a new world order, are far from over. During 1995 the right-wing politicians in the United States who had regained powerful positions in Congress mounted a strong assault on the whole idea of arms control and the progress that had been made over the decade since 1985. They focused on attempts to 'ravage' the central Anti-Ballistic Missile Treaty and on blocking ratification of START II[13,14] but it was also reported that the chairman of the crucial Foreign Relations Committee, Senator Helms:[15]

> . . . and many other senators are opposed to the Chemical Weapons
> Convention because of serious concerns about verification . . .

The battle over the future of arms control seemed set to continue in the United States and this was bound to be observed, and to have potentially serious repercussions, in other states such as Russia.

From our viewpoint, however, it is important to repeat that some percipient analysts are beginning to argue that there may be another military-technology revolution underway – that of non-lethal weapons specifically designed for purposes of intervention by advanced industrialised states. It is feared that the undoubted advantages of new technology in conventional combined arms conflict will lead to an unthinking attempt to apply new technologies to conflicts short of war where the advantages may turn out to be fewer than suggested in some optimistic scenarios now being advanced by supporters of 'benign interventions'.

Steven Metz, of the US Army War College, has attempted to focus thinking about the possible consequences of moving in this direction by constructing a future scenario. He describes a hypothetical history, written in the year 2010, of the application of the revolution in military technology to conflicts short of war. In this history a series of fiascos in the mid-1990s led to the adoption of radically new methods which were eminently successful initially. But the technologies could not be contained and thus:[16]

> In 2010, a decade of constant success in counter-terrorism was marred by several dramatic failures. The post-attack environmental clean-up and reconstitution of St. Louis will take decades . . .

and:

> . . . Many of the difficult-to-detect drugs and psychotechnology developed for use in conflict short of war have appeared on the domestic black market. . . . Perhaps most important, Americans are beginning to question the economic, human, and ethical costs of our new strategy . . .

More recently, Metz has argued for a detailed study of the roots of insurgency in the new world system in order that appropriate responses may be developed.[17] Given the wide range of weapon systems envisaged for 'benign' interventions, no one mechanism of arms control could possibly be adequate.[18] A prior question, however, is whether the international law of arms control is in any way up to the task of controlling worldwide militarisation.

International Law and International Relations

There were, of course, American critics of the distancing of the discipline of international law from international relations during the Cold War period,[19] but Kenneth Abbott was surely broadly correct when he argued in 1992, at the Annual Meeting of the American Society of International Law, that:[20]

International law (IL) and international relations (IR) theory have been estranged for many years. This estrangement has been so complete as to be truly remarkable. . . . It has been difficult to tell for the last 20 years that the two disciplines were even talking about the same world.

The situation has changed rapidly within the field of international relations with quite hard-line realists writing papers with such titles as 'Realists as optimists: Co-operation as self-help' and arguing that arms control – at least in some circumstances – is a reasonable course of action.[21]

The 'English' school has always had an important place for co-operation and law in an anarchic world, particularly in the work of Hedley Bull. As Buzan has argued in his attempt to link the 'English' school with US structural realism:[22]

> . . . Units that have no choice but to interact with each other on a regular long-term basis, and that begin to accept each other as essentially similar types of sociopolitical organisation, will be hard put to avoid creating some mechanisms for dealing with each other peacefully . . .

And as he stressed:

> . . . With the foundation of legal equality, much scope opens for the development of law as a way of ordering relations among sovereign states, though it can only develop where consensus allows.

There will be cause to return to Buzan's analysis, in regard to the building of consensus and regime theory, at the end of this section.

International lawyers are clearly in the middle of a major debate on the nature of their subject[23] and its relationship to international relations,[24] but it is striking that, in a study of 'organised international co-operation',[20] a leading figure such as Abbott saw advantages in merging the two disciplines. He went on to analyse what each might contribute to five different aspects of scholarship in the joint endeavour and one of his conclusions is of particular interest here. He argued that:

> . . . the kinds of complementary scholarly work already described – particularly analysis of the functions that international rules, regimes and institutions perform, how they are created . . . – could greatly strengthen our ability to perform such creative instrumental tasks as designing effective agreements, procedures and institutions . . .

In short, together these two disciplines may be in a position to contribute effectively to the job of controlling worldwide militarisation.

Yet it has to be asked what precisely is the nature of the task? Can we proceed with a series of essentially *ad hoc* agreements as and when political forces allow, or is something much more serious and systematic required? It seems that the militarisation processes which were accelerated in the Cold War period leave us with little choice. The increasing lethality of weaponry available to more and more states (and sub-state groups) means that the only effective solution to many arms control problems is to seek worldwide bans – **global prohibition regimes**. If this route is not taken in regard to hand-held, blinding laser weapons, for example, we may expect to repeat, in a decade or two, the disaster which has resulted from our failure to ban anti-personnel mines. Both weapon systems are cheap and neither employs technology of great sophistication. In fact, we have to relearn our history and realise that the dominance of bipolar nuclear arms control from 1960 to the end of the Cold War was preceded by a period in which statesmen who had experienced the devastation of the Second World War seriously attempted to find a means for General and Complete Disarmament.[25]

Because these efforts have to a large extent been forgotten during the 30-year dominance of a much narrower conception of arms control, it is difficult to grasp how extensive and thoroughgoing they were. All the plans put forward by both East and West envisaged a series of stages of disarmament covering both nuclear and conventional weapons and ending with very low levels of armed forces. In 1962, for example, following the McCloy–Zorin 'Joint Statement of Agreed Principles for Disarmament Negotiations', the United States presented a draft General and Complete Disarmament treaty at Geneva:

> The basic theme of the draft was an 'across-the-board' approach of one-third reductions [of the initial total] in the nuclear and conventional fields in each of the three stages. Each of the first two stages would be three years; the third stage would be completed 'as promptly as possible' . . .

The US and Soviet representatives in Geneva actually began an article-by-article analysis of the two sides' competing drafts and agreed some of the introductory text before the pressures of the Cold War, particularly the Cuban Missile Crisis, switched attention to more immediate measures such as the Hot Line and the Limited Test Ban Treaty.

Global prohibition regimes exist outside arms control, for example in the banning of piracy, slavery, hijacking of aircraft and trafficking in psychoactive drugs. While these regimes do reflect the interests of the most powerful states, Nadelmann is surely correct to argue that they also reflect widespread agreement on moral grounds.[26] Indeed, Nadelmann suggests that such global regimes generally follow a common evolutionary pattern of development through four or five stages. In the first stage the activity is widely regarded as legitimate; then in a second stage, often involving very active 'moral entrepreneurs', the activity comes to be viewed as illegitimate. In a third stage there is active agitation for the activity to be suppressed and criminalised. If this succeeds, a global prohibition regime comes into existence throughout much of the world (the fourth stage), and this can finally lead to a fifth stage in which the activity becomes insignificant, 'persisting only on a small scale and in obscure locations'.

It is perhaps worth repeating the conclusion of Nadelmann's analysis of the abolition of slavery. As he put it:

> . . . It is, in short, impossible to explain the abolition of slavery throughout the world during the past two centuries without emphasising the powerful role of shared moral notions derived primarily from the religious and secular principles of the Enlightenment.

We may be little further forward in our technical understanding of how value systems change than when Sir Geoffrey Vickers first suggested that we appreciate the world through the influence of our value systems rather than just perceive it, but we do know that changing value systems are a crucial component in effecting such huge changes in the international system.[27]

In his consideration of control regimes that may be developed, Nadelmann suggested that potential targets are 'the unauthorised development and distribution of atomic, biological, chemical and other weapons that can be used to wreak great destruction'. He argued that the possible spread of such weaponry to sub-state groups would force governments to co-operate in control measures. He clearly saw these pressures mounting and suggested, therefore, that the prospects for global prohibition regimes applying to the weapons were good. One reservation needs to be kept in mind, however. Effective global regimes require global consensus.

In the latter part of his essay Buzan[22] attempted to tie together his reworking of structural realism, the 'English' school, and regime theory. This helps to illuminate the basic problem of consensus. In Buzan's view, our present international society is partly formed by the organic European group that spread across the world and partly of the relationships they then constructed with the groups they encountered. Naturally, given this mode of development, global society:

> . . . has a European (now Western) core that is much more highly developed than the rest of it in terms of having a higher number, variety, and intensity of rules, norms and institutions binding its members in a network of regimes . . .

but:

> . . . as one would expect from its partly gesellschaft [artificially constructed society] origins, it is globally multicultural in character and significantly differentiated in terms of the degree of commitment with which states adhere to it . . .

So, whilst the end of the Cold War has led to a more cohesive global society, Buzan sees an outer circle of states with peripheral or very little commitment to international society (eg North Korea). More significantly:

> . . . In the middle circles one finds states such as Argentina, China and India that seek to preserve high levels of independence and select quite carefully what norms, rules and institutions they accept and what they reject . . .

Building consensus for global regimes requires that the inner core of states at least finds means of convincing significant states of Buzan's middle circle – such as India and China – that it is in their interests to participate. This is not, of course, to argue that consensus is the only requirement for a regime to develop[28] but it is surely an essential one.

The Chemical Weapons Convention

Increasingly, non-proliferation regimes are analysed as a group. Leonard Spector and Jonathan Dean, for example, have assessed the regimes controlling nuclear, biological, chemical and missile proliferation according to the co-operative security design criteria of: normative base; inclusiveness and non-discrimination; transparency; regime management; and sanctions. When it comes into force, the Chemical Weapons Convention (CWC) will be the strongest of the regimes on such criteria.[29] With these authors accepting a definition of co-operative security as:

> . . . in essence, a commitment to regulate the size, technical composition, investment patterns, and operational practices of all military forces by mutual consent for mutual benefit . . .

it is reasonable to argue that the CWC must contain many of the elements that it will be important to incorporate in strong regimes in the future.

The CWC was eventually agreed in the early 1990s after the ending of the Cold War, and following detailed and protracted negotiations at the Conference on Disarmament (CD) in Geneva through most of the 1980s.[30] It should also be remembered that it took almost 100 years from the initial recognition of the potential problem of the use of chemical weapons in modern warfare, at The Hague Peace Conferences at the turn of the last century, to the final agreement of the CWC. At various critical points during that time, it was by no means certain that the international norm – that these were unacceptable weapons – could be maintained and a strong global prohibition regime developed.[31]

The Chemical Weapons Convention was opened for signature in January 1993. It will enter into force 180 days after 65 states have ratified their commitment. It is hoped that this will take place in 1996. The massive convention text consists of a preamble, 24 articles (Table 9.1) and three annexes.[32] All the parties to the Convention will undertake never to 'develop, produce, otherwise acquire, stockpile or retain chemical weapons' or to use such weapons or assist others to engage in prohibited activities. They will also undertake to destroy, within ten years of entry into force, their chemical weapons, production facilities which can produce significant quantities of chemical weapons, and chemical weapons abandoned on the territory of another party. With the US possessing some 30,000 tons of agent and Russia 40,000 tons this, in itself, will be a massive undertaking.[33]

The Convention (Article 2) has a very wide definition of the toxic chemicals of concern:

> Any chemical which through its chemical action on life processes can cause death, temporary incapacitation or permanent harm to humans or animals . . .

However, the same article also lists purposes which are not prohibited such as 'Industrial, agricultural, research, medical, pharmaceutical and other peaceful purposes', as well as protective purposes, military uses not connected with the use of chemical weapons and, interestingly, as we shall see later, 'law enforcement including domestic riot control purposes'.

Table 9.1   Outline of the Articles of the Chemical Weapons Convention (CWC)*

Preamble	
Article 1.	General Obligations
Article 2.	Definitions and Criteria
Article 3.	Declarations
Article 4.	Chemical Weapons
Article 5.	Chemical Weapons Production Facilities
Article 6.	Activities not Prohibited under this Convention
Article 7.	National Implementation Measures
Article 8.	The Organisation
Article 9.	Consultation, Co-operation and Fact-Finding
Article 10.	Assistance and Protection against Chemical Weapons
Article 11.	Economic and Technological Development
Article 12.	Measures to Redress a Situation and to Ensure Compliance, Including Sanctions
Article 13.	Relation to Other International Agreements
Article 14.	Settlement of Disputes
Article 15.	Amendments
Article 16.	Duration and Withdrawal
Article 17.	Status of the Annexes
Articles 18–24.	Signature, Ratification, Accession, Entry into Force, Reservations, Depository and Authentic Texts
Annexes	

* From reference 32

To guarantee the operation of the Convention a large Organisation for the Prohibition of Chemical Weapons is to be established, with its headquarters in The Hague. The organisation will consist of a high-level Conference of the State Parties which will meet once a year, an Executive Council of 41 carefully-balanced States Parties, and a Technical Secretariat which will be responsible for day-to-day activities. At the time of writing, a Provisional Secretariat is carrying out the work necessary prior to the Convention entering into force. In particular, the Convention will require a complex system of declarations, and routine and challenge inspections, to verify that the specific chemicals of concern are not being used for prohibited purposes. These chemicals are listed in three schedules with declining stringency of control as the lethality of the toxic chemicals declines. Schedule I, for example, lists all the nerve and mustard gases, and effectively bans them from manufacture and trade. It will be noted from Articles 10, 12 and 14 that the Convention allows for tough action to be taken in the event of non-compliance. Both the Conference and the Executive will be able to take issues of grave concern to the Security Council of the United Nations.

The CWC has attracted a great deal of attention in the international relations/security studies literature. Yet, as Abbott pointed out in his assessment of how international relations and international law scholars might effectively co-operate:[20]

> ... IR is heavy on theory, but relatively light on empirical application and testing. IL can bring to the table a wealth of data on legal practices and procedures, primary rules of conduct, secondary rules of law formation, interpretation and application ...

What then do international lawyers make of the CWC?

Barry Kellman is an American lawyer who chaired the team of international experts who prepared the 'Manual for National Implementation of the Chemical Weapons Convention' for the Organisation for the Prohibition of Chemical Weapons. He has written at length recently on the problem of preventing the trade in what he calls 'catastrophic weapons'. These are nuclear, biological, chemical and missile systems whose use signals, in his opinion, a move from limited to total warfare.[34] In general terms, Kellman argued that, in a traditional *realpolitik* approach, the crux of the problem is seen to be the difficulty of deterring the use of such weapons. He argues, instead, for a 'regulatory' approach in which the essence of the problem is seen to be the prevention of suppliers from spreading the weaponry to other countries.

At a deeper level Kellman sees the *realpolitik* approach as having the goal of preserving security through hegemony and alliances, thus risking the continuation of the arms race in these dangerous weapons that has led to their proliferation over the past decades. He argues, on the other hand, that the non-discriminatory, regulatory approach opens up the possibility of clamping down on the proliferation of these weapon systems by building international arms control regimes. In his words:[34]

> ... a regime propounds standards of conduct to which state parties must accede. While each nation retains the option to join a regime, it has little freedom to adapt a regime to its particular interests and must accept the uniform standards that apply to every other participant ...

In his detailed analysis of the four regimes presently in place for catastrophic weaponry he shows that what we have is a patchwork of different controls, with even the CWC lacking in certain critical elements. He therefore sets out what he sees as necessary elements for all regimes in this area (Table 9.2). In Kellman's view the last two elements (5c and 5d), involving the prescription of transnational law that states may use to investigate suspicions and the ability to impose sanctions, are necessary, radical, new means that will give regimes the legal force they require.

Table 9.2 Outline of a comprehensive non-proliferation regime*

---

1. Aim to regulate the supply of critical military technology.

2. Add to, not abandon, what is now in place.

3. Use principles of industrial and trade regulation.

4. Establish institutional mechanisms that limit availability of critical technologies by:

    a) Detailed licensing requirements that can be objectively set out and verified;

    b) Enforcement mechanisms with specified penalties for proliferators.

5. The regime to have four essential functions:

    a) Formulation of the standards that uniformly restrict the production and trade of critical items;

    b) Development of a verification system complex enough to assure compliance and equality of burden-sharing;

    c) Prescription of the means of transnational law that states can use to investigate suspicions of potential proliferation;

    d) Ability to impose penalties on nations, economic entities and individuals involved in proliferation.

6. Functions 5c and 5d are radical developments necessary to close the gaps in current control mechanisms.

---

* From reference 34

Drawing on his extensive analysis of existing regimes, Kellman suggests that, in regard to technology controls (5a, Table 9.2), a regime has to begin by creating a graduated series of schedules with the extent of regulation depending on the importance of the particular schedule. The kinds of question that need to be asked when judging the appropriate schedule are, in his opinion, the importance of an item in making a weapon (taking account of possible substitutes) and whether the item has other possible uses. Items that are exclusively of use for weapons-making should be subject to intense control, and production should be prohibited except in exceptional circumstances. Dual-use items, on the other hand, should be equally available to States Parties, but export restrictions should again be linked to the importance of items in weapons capability. Kellman argues strongly that the aim should be to minimise the number of items controlled. 'Higher fences around fewer goods' is the mechanism he suggests as most appropriate. Nevertheless, there are undoubtedly difficult problems, for example in dealing with items just below agreed specifications and, more generally, in dealing with new technology as it arises. Despite this, he believes that a strong regime should require that end-users are declared and that these destinations may be verified. Moreover, access to critical technologies should be denied to those outside the regime or those found not to be complying with it.

Regarding verification (5b in Table 9.2), Kellman argues that the aim now must be to detect whether a state is going beyond the regime's limits in a militarily significant manner. This should not be confused with verification during the Cold War between the superpowers. Then, war avoidance was predicated on strategic nuclear stability, and any break-out could have been fraught with catastrophic consequences. Thus attention was focused on force structure and capability, both for military reasons and because any cheating could have threatened the core political relationship.

Nowadays, small discrepancies are not of great individual concern because a multilayered verification system can pick up broader, militarily significant evasion in good time. Indeed, at least in regard to chemical and biological weapons, it is unlikely that any verification system could assure *absolute* non-production because such a wide range of people and facilities could be used for some manufacture. The building of a significant arsenal, however, should be detectable. In Kellman's opinion, therefore, the target of verification now must be industrial capability, not deployed force. Early, intrusive verification measures for commercial enterprises will ensure that a single break-out effort cannot upset strategic balances. Kellman believes the CWC's verification system is what should be required in all regimes, in order to make the necessary assessment of production capabilities. For exportable items, a system of manifests which allows a traceable paper trail is required.

Turning to the radically new measures, Kellman argues that the lack of well-developed means of law enforcement has weakened all non-proliferation regimes to date (5c in Table 9.2). He suggests that:

> The most important single step that can be taken to control weapons proliferation would be to treat it as a penal, rather than diplomatic, matter. If proliferation is a crime, then it would no longer be diplomatically acceptable to sell weapons . . .

This would transform the present situation, where a primary justification for proliferation is that everyone else may be doing it, to one in which non-proliferation would be the norm because intense scrutiny would uncover evasion. Nevertheless, it is to be expected that some states would still attempt evasion by segmenting production of controlled items. Given the political difficulties of demanding challenge inspections of another state on the basis of suspicion of such activity, and the difficulties of achieving effective co-operation between courts in different countries, Kellman argues that a regime should include formal mechanisms of legal co-operation between states:

> A state party should be obligated to provide information relating to . . . suspect activities of persons or entities within its jurisdiction, including disclosing information it possesses, conducting searches, examining sites or objects, taking evidence or statements from persons, effecting service of judicial papers, and providing relevant documents . . .

As we noted in Chapter 7, the UN Convention Against Illicit Traffic in Narcotic Drugs and Psychotropic Substances already requires a state to provide such assistance without regard to domestic banking secrecy laws.

In Kellman's view there will need to be a national authority in each State Party which will co-ordinate the required regime activities. These would include gathering of information, licensing of dual-use exports, expediting of on-site monitoring and law enforcement (for which it would co-operate direct with other national authorities). This focused, systematic concentration of effort would contrast sharply with the present distributed responsibility in many departments of state.

Finally, in regard to penalties (5d in Table 9.2) Kellman argues:

> The absence of any formal statement of consequences for proliferation is the single most serious weakness of current control efforts. Accordingly, the single most important step that a new non-proliferation regime could take would be to promulgate specific diplomatic and trade penalties for . . . entities that disobey the regime's technology restrictions . . .

Since the pursuit of profit is now the main motivation driving proliferation, he argues that the most effective sanction could be economic. In particular, he argues that imposition of import controls against goods from the offending state or private entity, raised multilaterally within the international community, would be an effective coercive measure and would not greatly harm the states taking this corrective action.

All these features of strong regimes are based on co-operative, multilateral, consensual agreements, but even with the CWC, which is presently the most advanced, fear and suspicion both of discrimination and of imbalance of rights and duties persist.[35] It is therefore important that Western states do not, even inadvertently, cause problems by appearing to manipulate regime agreements for their own purposes. A particular example arises in this regard in relation to riot control agents under the CWC.

The drive to produce non-lethal weapons has caused concern because of the potential impact on a number of arms control agreements.[36] Most fundamentally, international humanitarian law has clearly developed a norm against weapons which could cause superfluous injury or have indiscriminate effects. As we have seen, the introduction of hand-held blinding lasers would necessarily cause concern on these grounds. Similarly, the use of bacteria to degrade fuel would appear to be directly contrary to the Biological Weapons Convention. The question of riot control agents within the CWC is therefore a specific instance of a general problem.

In its review of the successful conclusion of the CWC negotiations, the *SIPRI Yearbook* of 1993 pointed out that a compromise had been needed on the question of riot control agents.[30] The authors commented:

> Riot control agents were a problem throughout the history of the negotiations. While largely used for law enforcement and crowd control . . . they could constitute a risk to the CWC if developed into a new generation of non-lethal, but effective, warfare agents.

They concluded that this issue would require careful monitoring.

To understand what is at stake we need to look a little more closely at the CWC. Extracts of the most relevant Articles are set out in Table 9.3. As we have seen, each state undertakes in Article 1.1 not to develop chemical weapons. Chemical weapons are defined very widely in Article 2.1(a) to include all toxic chemicals except those intended for purposes that are not prohibited by the Convention. This 'general purpose criterion' means that new chemicals developed through scientific progress would still be covered by the Convention. Article 2.9 then sets out the purposes which are not prohibited. These include 2.9(d), 'Law enforcement including domestic riot control purposes'.

Table 9.3   Relevant extracts from Articles of the Chemical Weapons Convention

---

Article 1. General Obligations

    1.1 Each State Party . . . undertakes never . . .

        (a)   To develop, produce, otherwise acquire, stockpile or retain chemical weapons . . .

    1.5 Each State Party undertakes not to use riot control agents as a method of warfare.

Article 2. Definitions and Criteria

    2.1 'Chemical Weapons' means . . .

        (a)   Toxic chemicals and their precursors, except where intended for purposes not prohibited . . .

    2.2 'Toxic Chemical' means:

    Any chemical which through its chemical action on life processes can cause death, temporary incapacitation or permanent harm to humans or animals . . .

    2.7 'Riot Control Agent' means:

    Any chemical not listed in a Schedule, which can produce rapidly in humans sensory irritation or disabling physical effects which disappear within a short time following termination of exposure.

    2.9 'Purposes Not Prohibited . . . ' means:

        (d)   Law enforcement including domestic riot control purposes.

Article 3. Declarations

    3.1 Each State Party shall submit . . . declarations, in which it shall:

        (e)   With respect to riot control agents: Specify the chemical name, structural formula and Chemical Abstracts Service (CAS) registry number, if assigned, of each chemical it holds for riot control purposes. This declaration shall be updated not later than 30 days after any change becomes effective.

Now a riot control agent is clearly defined in Article 2.7 as:

Any chemical not listed in a Schedule, which can produce rapidly in humans sensory irritation or disabling physical effects which disappear within a short time following termination of exposure.

So it is quite clear that the use of tear gas for domestic riot control is allowed by the CWC, but Article 3.1(e) makes specific declarations of the possession and identity of such agents mandatory.

The first obvious complication concerns 'law enforcement'. This is not defined in the Convention and the only reference to chemicals for law enforcement occurs in the Verification Annexe, where it is stated that Schedule 1 chemicals (the most dangerous) may not be used for law enforcement. The failure to define law enforcement, or law enforcement chemicals, creates ambiguity as to the meaning of the Convention. One commentary asked:[37]

> ... Is the Convention really to be read as allowing any non-Schedule-1 toxic chemical or precursor to be developed, produced, weaponised, stockpiled or traded, so long as it is said to be for 'law enforcement purposes'?

Obviously, one would think that was not what was intended, because it would open a huge loophole allowing the development of new, undisclosed chemicals and thus arouse grave suspicions among State Parties. Worryingly, the negotiation record suggests that:[37]

> ... Some, by no means a majority, of the negotiating states wished to protect possible applications of disabling chemicals that would either go beyond, or might be criticised as going beyond, applications hitherto customary in the hands of domestic police forces ...

This seems to imply that the ambiguity in the CWC did not come about by accident, but was deliberately intended by some parties to the negotiation. The United States, as we have seen, has for some time been developing an Advanced Riot Control Agent Device (ARCAD) intended to 'deliver a high safety ratio immobilisation compound against relatively close targets where safe immobilisation is the prime concern'.[38]

A further complication arises from Article 1.5 which states that 'Each State Party undertakes not to use riot control agents as a method of warfare'. Given that riot control agents have been used extensively as a method of warfare in the past, this might be read as an extra reinforcement of the purpose of the Convention. Again, however, the precise meaning of 'method of warfare' is not given. The US interpretation is that while the use of riot control agents is prohibited in international and internal armed conflict, use in:[39]

> ... normal peacekeeping operations, law enforcement operations, humanitarian and disaster relief operations, counter-terrorist and hostage rescue operations and non-combat rescue operations ...

will be allowed if conducted outwith inter- and intra-state conflicts. Furthermore, the United States does not even consider that the CWC applies to all uses of RCAs during armed conflict. Use of such substances solely against non-combatants would be allowed, for example:

> . . . in riot control situations in areas under direct US military control including against rioting prisoners of war, and to protect convoys from civil disturbances, terrorists, and paramilitary organisations in rear areas outside the zone of immediate combat.

Finally, while the new CWC is understood to prevent the use of RCAs against combatants (even if mingled with non-combatants), 'were the international understanding of this issue to change, the United States would not consider itself bound by this position'. Reports that three varieties of pepper spray and a suds-like barrier laced with tear gas were selected for possible use by US Marines in the evacuation of Somalia are understandable in this context.[40]

The immediate danger is that incorporation of such weapons systems into the complex type of peacekeeping operations that are likely to occur frequently in the future would complicate control for commanders on the ground.[41] This would be particularly true for troops not well versed in the need for very special attention to consent in Article VI operations. In the longer term, it is easy to see why a break-out by advanced industrial states from the CWC's general purpose ban in this area could lead to an erosion of the norm, which forbids use of such chemicals, among other states. If these chemicals are seen as advantageous by the military forces of the US, why should they not be seen in the same light by others? Again the general problem is that immediate short-term advantages conferred by technological developments could endanger one prohibition regime and ultimately the complete set of arms control regimes that the international community is trying to erect to restrain the proliferation of advanced weaponry. It is the institutional underpinning of that effort that we shall now consider.

Institutions and Arms Control

Although thinking about arms control is usually compartmentalised in relation to different major areas of concern and the treaties relevant to those areas, the wide range of non-lethal weapons under development that might be used in future interventions forces a broader approach. Any broader conceptualisation of arms control and the mechanisms for creating and operating arms agreements necessarily leads to consideration of the United Nations system.

The roles of the General Assembly, the First Committee and the Disarmament Commission in disarmament debates are widely known. What is perhaps less well understood is that, while the UN Charter had much less to say on the subject than the Covenant of the League of Nations, the 'Security Council was to formulate plans for the establishment of a "system" for the regulation of armaments'. This it indeed began to do during the early post-war period, and the Military Staff Committee of the UN was charged with providing necessary assistance.[42] It seems likely that we shall eventually see arms control and disarmament guided from that level, with a coherent overall policy developed by a reformed Council with enhanced legitimacy.[43]

To those who understandably take the view that the UN has had little success in developing an international legal system for the regulation of armaments, it is perhaps necessary to restate that, while the founders of the UN may not have intended it, the UN has created a vast network of international law by a wide variety of mechanisms, despite the immobilizing effect of the Cold War.[44] Arms control has also, in fact, been a major area of activity. In the new era, when multilateral negotiations are likely to increase in importance, it is probable that the Conference on Disarmament in Geneva will become increasingly prominent as the main negotiating body in the system.[27] While efforts to reform its membership, agenda and procedures may be taking longer than some had hoped, it seems almost certain that these matters will be resolved as the necessary concentration on the Nuclear Non-Proliferation Treaty declines following the completion of its 25-year Review Conference.

Though the standard issues of nuclear and conventional disarmament are going to be part of any reassessment, as Bertrand's analysis would lead us to expect,[7] the new problems of peacekeeping are beginning to provoke a rethinking of what arms control arrangements are required. Klaus Törnudd, a Finnish member of the Secretary General's Advisory Board on Disarmament, has written:[45]

> One of the new key ideas is that disarmament should be *integrated* with other United Nations activities in the field of peace and security . . .

He suggests that 'preventive diplomacy' – efforts to prevent difficult conflict situations from deteriorating – could be assisted by 'preventive disarmament'. He argues, for example, that it should be possible to develop a standard repertoire of military confidence-building measures to use at various stages of a preventive action. Similarly, he briefly describes ways in which arms control and disarmament measures might be effectively integrated in peacemaking, peacekeeping, peace enforcement and peace-building.

At an institutional level Törnudd argues that:

> . . . there should be close liaison and cooperation within the UN Secretariat between the units responsible for disarmament affairs and peacekeeping, respectively. In the preparation of mandates for peacekeeping operations, the inclusion of tasks pertaining to disarmament in the broadest sense should always be considered.

It is significant, as the UN Secretary General has commented,[46] that the First Committee is already combining 'consideration of arms control and disarmament with the wider concerns of international security'. It seems probable that this process will accelerate. The Swedish Prime Minister, for example, reporting on the 1995 findings of his 'Commission on Global Governance', stressed the important role of the UN in disarmament and the necessity of linking savings from declining arms budgets to economic and social development.[47] What we may be seeing is an interaction of norm creation and institution development which will increasingly elevate longer-term security considerations above short-term military advantage.

. 10 .

BENIGN INTERVENTIONS OR A
NEW ARMS RACE?

Late in 1995 it appeared that the vocal and influential lobby for the systematic incorporation of non-lethal weapons into the armoury of the US military had won significant victories. *Defense Week*[1] reported that the Pentagon was just completing a plan to guide research and development on the most promising non-lethal technologies. More importantly, this was one of three initiatives concerned with non-lethal weaponry. According to the report:

> Together, the initiatives constitute a milestone marking the Pentagon's apparent advancement on non-lethal technology from theory to doctrine and materiel.

This is made patently clear by the fact that the other two initiatives were a draft policy document on non-lethal weapons issued in June 1995 and acquisition plans for non-lethal weapons.

Before looking at some of these plans, the context within which non-lethal weapons are being considered by industrialised states such as the United States will be outlined. We shall also reconsider the wide range of technologies involved, before attempting to review the various arguments being advanced to support, question, or oppose the movement to systematically incorporate non-lethal concepts and weapons into modern military strategy and operations.

A World in Transition

The International Institute for Strategic Studies in London opened its

191

Strategic Survey for 1994–95 in pessimistic tone:[2]

> A pervading sense of impotence characterised international affairs during the past year. As a consequence it was a period of drift. . . . The world appeared to be marking time, while many of the positive advances that had been made since the end of the bipolar geopolitics that had dominated the past 45 years receded . . .

The world system, the Institute argued, was in a period of transition from the bipolar period of Cold War confrontation, and the formulation and implementation of coherent foreign and defence policies in such a context were not straightforward. Yet there was often broad agreement emerging, in analyses of the nature of that context, from northern industrialised nations.

The Institute for National Strategic Studies of the US National Defense University, for example, carried out a comprehensive survey of the 'emerging strategic landscape', which was summarised in an article by Hans Binnendijk and Patrick Clawson in *The Washington Quarterly*.[3] They clearly saw the transitional world system remaining a divided one for some time to come, with most of the probable conflicts arising in poorer, unstable or failed states (Table 10.1). Priorities for the US military, therefore, necessarily involved preparing for engagement in regional conflicts, responding to transnational threats and assisting failed states. While problems at the last of these levels are the least likely to involve US vital interests directly they are certainly taken seriously, since repetitions of the debâcle in Somalia would be seen as threatening both the credibility of and the will to undertake other, possibly much more important, such missions in the future.

Another prominent and recurrent feature of discussions of future security remains the emphasis placed on the revolutionary nature of modern military technology. As Admiral Owens, Vice Chairman of the US Joint Chiefs of Staff, remarked:[4]

> If we decide to accelerate the process by emphasising those systems and weapons that drive the revolution, we can reach our goals years – perhaps decades – before any other nation . . .

He continued:

> . . . If we can combine the new military capabilities the transition will give us with foreign and security policies that are consistent with those capabilities, we will be able to influence the character of the international system . . .

Table 10.1   The emerging strategic landscape *

The most likely conflicts to be expected in the emerging world system are the least dangerous to the United States

- Conflict among the major powers
- Conflict among regional powers
- Conflict involving troubled states.

Proposed priorities for US security policy

- Ensuring peace among the major powers
- Engaging selectively in regional conflicts
- Responding to transnational threats
- Assisting failed states.

Implications for the US military

- Balance forces among four fundamentally different missions:
    - Hedging against the emergence in the next one or two decades of a military peer competitor from among the major powers
    - Preparing for major regional conflicts with rogue states
    - Developing a cost-effective response for quasi-police missions in order to meet transnational threats
    - Engaging selectively in troubled states.

Conclusion

'The last priority item is peace operations to deal with failing states. This is the most common new threat but also the least likely to affect US vital interests . . .'

'Operations like that in Somalia can erode both the credibility of the United States and its will to take determined action elsewhere where it can be more effective at promoting its humanitarian and democratic values as well as its national interests . . .'

---

* From reference 3

In such a general context, advocates of novel non-lethal weaponry – arguably appropriate for quasi-police and low-level conflict operations and relying on new technology – were clearly in an advantageous position to pursue their case.

Policy Developments in Washington

In July 1995 *Inside the Pentagon*[5] reported on the draft policy directive on non-lethal weapons issued in late June and suggested that non-lethal weapons were important, 'because they further enhance US military superiority'. The draft document itself defined non-lethal weapons as:[6]

> . . . weapons that are explicitly designed and employed so as to incapacitate personnel or material, while minimising fatalities and undesired damage to property and the environment.

To allow maximum flexibility for the employment of US forces across the full range of military operations, the draft stated that additional non-lethal options would be developed. These new options would clearly contribute to achieving the aims set out in Table 10.2.

Table 10.2 Aims for non-lethal weapons policy*

- Discourage, delay or prevent hostile actions by prospective opponents;
- Limit escalation;
- Take military action in situations where intervention is desirable but use of lethal force would be inappropriate;
- Better protect our forces once deployed;
- Provide an effective but also reversible and more humanitarian means of denying an enemy the use of some of his human and material assets;
- Reduce the post-conflict costs of rebuilding infrastructure.

---

* From reference 6

The draft document clarified a number of crucial points about the way that non-lethal options are regarded, and would be used, by the Pentagon. Firstly, the military technological advantage is made quite clear. Non-lethal weapons would make:

> . . . potential adversaries aware that the United States can thwart aggression and achieve humanitarian aims in ways that do not entail excessive costs, thus enabling us to act earlier, more freely, and more decisively.

Secondly, while the weapons are described as non-lethal:

> When employed, some non-lethal weapons may produce fatalities, either because of imperfect control over all factors, or because they are used in conjunction with lethal weapons in order to more quickly terminate conflict and reduce overall suffering . . .

It must be said that similar points have frequently been made by some of the main public advocates of non-lethal weaponry in the United States.[7] The draft document also makes it clear that non-lethal weapons are not regarded as a means of supplanting conventional warfare nor will they consume large amounts of the defence budget. Nevertheless, certain capabilities will be prioritised and sought (Table 10.3), should the policy be implemented.

Table 10.3 Priority non-lethal capabilities sought*

- Disarm combatants intermingled with non-combatants;
- Control crowds;
- Support security of facilities and our own forces;
- Incapacitate tactical civil and military logistics;
- Disable elements of, or the entirety of, regional civil/military communication, transportation, and energy infrastructures;
- Incapacitate/immobilise weapons or weapon development/production processes, including suspected weapons of mass destruction;
- Temporarily incapacitate tactical forces.

---

* From reference 6

The draft policy document makes clear that the tasks set out in Table 10.3 could become necessary anywhere across the range of military operations. An Army document, 'Draft concept for non-lethal capabilities in army operations', of August 1995, helps to clarify the potential application of non-lethal weapons.[8] A list of the capabilities mentioned in this Army Training and Development Command document is given in Table 10.4 with some illustrative examples.

While to the non-advocate this may still seem to be a motley collection of capabilities and operations for consideration under a single heading, there is no doubt that it is persuasive to budget controllers. The Senate Committee on Armed Services remarked:[9]

> The National Defense Authorisation Act for Fiscal Year 1995 authorised $41.0 million for work on non-lethal weapons technology applicable to peacekeeping and law enforcement. The committee supports the continuation of the efforts to identify, evaluate, develop, and field non-lethal systems and technologies, and recommends $37.2 million for fiscal year 1996 . . .

The Committee went on to praise the Marine Corps for validating the 'operational utility' of non-lethal weapons in Somalia but, because of the significant shortcomings revealed in the Marine Corps operation, it also directed the consolidation of the non-lethal programme under one

Table 10.4   Operational employment of non-lethal weapons*

---

(1)   Offensive measures

  (a)   Riot/Mob Control

  '[Non-lethal weapons] could be used to disperse dangerous mobs or deny access to such critical facilities as weapon storage sites, embassies . . .'

  (b)   Sanctions

  '. . . Economic blockades could be improved by employing measures that could rapidly degrade vehicle tyres and lines of communication . . .'

  (c)   Interdiction of Tactical/Strategic Resources

  'An important consideration in the future will be the ability to employ measures that degrade an adversary's capability to wage war, yet permit the reestablishment of the infrastructure with minimum costs . . .'

  (d)   Conflict Intervention

  'Non-lethal capabilities could be employed pre-emptively before the onset of hostilities or covertly after initiation of open conflict between two countries . . .'

  (e)   Military Incursions

  'US forces may be required to go into a country to accomplish a single objective, such as destruction of chemical production facilities or the capture of nuclear weapons . . .'

  (f)   Counterdrug/Terrorist Operations

  '. . . Non-lethal capabilities could include the emplacement of unattended sensors designed to detect use of aircraft or vehicles . . .'

  (g)   Hostage Retrieval

  'This use of non-lethal capabilities could enable the neutralisation of combatants when intermingled with non-combatants or friendly forces . . .'

  (h)   Military Operations in Urban Terrain

  '. . . Access/escape routes can be blocked in buildings . . . to . . . channel movements through established firing zones, or protect areas from entry . . .'

  (i)   Large-Scale Operations

  '. . . Strategic interdiction of warmaking necessities such as electricity and petroleum, oil, and lubricant resources could hasten the end of conflict . . .'

(2)   Defensive Measures

  '. . . Non-lethal capabilities could be employed to limit access to defensive positions . . .'

(3)   Non-lethal Countermeasures

  '. . . The Army must possess and employ appropriate countermeasures (especially for mission critical resources) to an adversary's use of non-lethal weapons/systems.'

---

* From reference 8

office. This consolidated Program Office for Non-lethal Systems and Technology was also to have responsibility for:

> . . . other funds being used to support highly classified programs in non-lethal technology and Operations Other Than War . . .

The true extent of US spending on these systems is therefore a matter of speculation.

What is very clear, however, is that development of new systems is continuing. The Advanced Research Projects Agency of the DoD sought proposals for new projects in May 1995.[10] Under the heading of Limited Effects Technologies, it was letting $3.2 million of contracts, for example for alternatives to 'contain hostile crowds, mobs, and rioters in urban environments with minimal potential for injury to belligerents or bystanders, or damage to property'.

Evaluation

It seems likely that development of new non-lethal weapons will continue for some years to come in the United States. Moreover, the US documents cited make clear the official view that many other countries are proceeding down the same path (hence the need to study countermeasures). So we cannot dodge the issue of evaluation: these new weapons will arrive and will have consequences in the real world of armed forces deployed in conflicts. What should we do? Should we welcome the development or should we do what we can to limit it?

On one point we can all surely agree. Because we are unlikely, in the near future, to find a way of preventing warfare, any means of effectively limiting the concomitant damage is to be welcomed. But what we need is a rounded evaluation, not a reflex response which says that because something is 'non-lethal' it must be good. As Richard Garwin, Vice-Chairman of the Federation of American Scientists, has pointed out, even land mines might be viewed as non-lethal weapons:[11]

> . . . An effective minefield prevents passage, and it can be argued that land mines put the initiative of injury on the other side . . .

With the experience of the last two decades, that particular case will obviously not stand much scrutiny, but it should alert us to the dangers which may be concealed in new proposals.

The difficulties of making a proper assessment are naturally compounded by the amazing variety of technologies and operational missions envisaged in proposals for new non-lethal weapons. How, for instance, are we to make any general statement which covers both a steel dart loaded with fentanyl which is said to be close to field use,[11] and proposals for the future massive use of non-lethal air power[12] to effect strategic paralysis of a country? A further complicating factor is the enthusiasm of proponents of non-lethal weaponry. There are now suggestions of a new phenomenon, 'weapons of mass protection'. As the Morris team suggested recently:[13]

> Acquiring weapons of mass protection – non-lethal, antilethal, and information warfare weapons – and integrating them into current force capabilities may be one way that airpower can secure for years to come its primacy in strategic utility for the post-cold-war conflict environment . . .

The best approach, perhaps, is to review briefly the general arguments for and against non-lethal weapons and then to look more closely at one or two specific examples to see how these arguments stand up in face of the facts. From these specific cases we may then be able to draw some conclusions which are worth extending to others.

The US Army concept document referred to previously[8] contains a brief summary of the reasons why non-lethal weaponry is needed. These five sets of reasons are set out in some detail in Table 10.5. In this official view, it seems that non-lethal weapons provide extra options for US forces, a point reinforced by a review of the 'background'. This section notes that non-lethal means have always been available to the military, including such 'classic' non-lethal means as:

> . . . show of force; deliberately delivered information or propaganda . . . physical obstacles; noise to create or enhance psychological effects; electromagnetic energy to disrupt communications; smoke and obscurants to mask operations or defeat homing and guidance mechanisms; and light or fires used to harass soldiers . . .

In this view, therefore, new non-lethal weapons derived from modern technology merely add to the already available list of possibilities.

Critics have argued that the overselling of non-lethal weapons in some quarters has allowed a variety of technology projects to survive and prosper without any critical attention being paid to their potential drawbacks.[14,15] Both critics and advocates are concerned about what else may emerge from secretly-funded ('black budget') programmes if the idea really takes hold. More mainstream analysts, such as those in the Independent Task Force set up by the Washington-based Council on

Table 10.5 Arguments for non-lethal weapons*

(a) '. . . US forces will have to respond to a myriad of military operations from war to OOTW [Operations Other Than War] . . .'

'. . . At the same time, the military will face increased media attention, worldwide environmental concerns, and a low national tolerance for long, lethal and costly campaigns. . .'

'. . . Non-lethal capabilities can expand options and tools available . . .'

(b) '. . . Non-lethal capabilities complement and extend the nation's diplomatic and military options beyond the use of more traditional lethal weapons . . .'

(c) '. . . Non-lethal capabilities support the objectives of thwarting aggression and promoting stability. . . . Recent operations . . . highlight the complexity and danger of missions across the range of military operations . . .'

(d) '. . . Non-lethal capabilities afford expanded crisis and contingency response options. . . . Non-lethal capabilities can reduce the risk of perceived excessive military force, promote international political support, alleviate environmental concerns, and enhance post-conflict transitions . . .'

(e) '. . . Potential adversaries have or are acquiring non-lethal capabilities . . .'

---

*From reference 8

Table 10.6 Arguments against non-lethal weapons*

The slippery slope
'The use of non-lethal weapons may seem an attractive option, but might lead to further unintended and unwanted involvement, including the large-scale use of lethal weapons.'

Retaliation
'Since the United States is highly dependent on technology, we may be opening the door to a form of warfare to which we are most vulnerable.'

Proliferation
'Much military research and development is based on mimicry. If we take the lead in developing non-lethal technologies, other countries will follow and renegades will eventually acquire them . . .'

Unrealistic expectations
'. . . An expectation of bloodless battles is doomed to disappointment . . .'

Comparative cost-effectiveness
'Many of the casualty-limiting benefits of non-lethal weapons could perhaps be achieved more quickly and at less cost by increasing the precision of lethal arms.'

Restraints of international law
'In some cases, the status of non-lethals is ambiguous under broadly drawn international conventions prohibiting the use of certain types of weapons or technologies.'

---

* From reference 16

Foreign Relations, whilst agreeing that the idea of using non-lethal means is worthy of close examination, have detailed a series of problems that need consideration.[16] These are set out in Table 10.6. The possible objections to non-lethal weapons include the following: that unrealistic expectations about such attractive options might lead to involvement in operations that deteriorate, as in Somalia recently; that by pushing forward with this new technology, advanced industrial countries may be developing ideal weapons to use against their own societies; that by demonstrating what can be done, these countries encourage others (perhaps less scrupulous) to do likewise; that moderating the effects of current lethal weapons by increasing precision may cost less in the long run; and that international treaties already achieved could be broken or, at least, appear to be threatened.

In short, there appears to be a formidable set of general considerations that must be taken into account when evaluating any new non-lethal option or the whole new set of non-lethal possibilities. It also has to be remembered that, in the United States[17] and other industrialised countries considering such options, defence budgets are declining and a real, internecine battle is under way between different elements of the Services over who will take on the remaining roles and missions. Historically, this has not been a fruitful situation in which to conduct rational analysis and debate. Bearing this in mind, we can now examine two specific cases: first, hand-held lasers, and second, incapacitating chemical agents.

It was reported, in July 1995, that the US Army had decided to proceed with production of a Laser Countermeasure System (LCMS) which could be mounted on a rifle and which was to provide 'the individual soldier with the capability to manually acquire and disable a threat electro-optic fire control system'.[18] Opponents of such systems, such as Human Rights Watch, argued that this decision should have been delayed until the results of the 1995 review of the Inhumane Weapons Convention were known, and Congressional representatives urged that the United States should support the proposed protocol, banning the use of lasers which intentionally blind, that was to be considered at the review.[19] It was clear to everyone involved that the military use of lasers for range-finding and target designation was becoming increasingly important.[20] But as Representative Lane Evans pointed out, in a letter to colleagues urging support for a ban:[19]

> . . . While the US may have an advantage in developing laser technology, following through and fielding these weapons would be a terribly ill-advised policy. While these weapons may give us a short-term edge on the battlefield, the long-term consequences for our soldiers would be devastating.

Failure to achieve an international measure of control of such weaponry, in short, would lead to its widespread deployment and, inevitably, its use against US personnel.

Close to the review conference, and responding to such pressure, the Secretary of Defense issued a statement on 1 September 1995:[21]

> The Department of Defense prohibits the use of lasers specifically designed to cause permanent blindness of unenhanced vision and supports negotiations prohibiting the use of such weapons . . .

Human Rights Watch welcomed this apparent change in US policy which, until then, had been to oppose totally any regulation of laser weapons. However, there appeared to be a loophole in the words 'unenhanced vision' which would allow the US to continue to develop all the lasers it wished:[22]

> The LCMS is characterised by the army as an anti-sensor or anti-optical system. However, when used against an individual employing an optical device, such as binoculars or gunner's sights, the LCMS would not destroy the optics, but rather would blind the individual as the optic magnifies the laser beam and focuses it on the most sensitive part of the eye . . .

The similarities in these events to the watering down of the original Inhumane Weapons Convention in the 1970s do not need to be emphasised. The balance of the argument is surely against producing such non-lethal weapons if an international agreement with real 'teeth' is possible. Unfortunately, as noted in the Preface, the end result of the 1995 review of the Inhumane Weapons Convention was the addition of a new protocol, designed to reflect the US position, which would allow deployment.

Regarding the second case, of incapacitant chemicals, there was also cause for concern over events in Washington in late summer and autumn 1995. Senator Helms, chairman of the Senate Foreign Relations Committee, was holding up ratification of the Chemical Weapons Convention, perhaps because he and others, in contradiction of US policy over the previous ten years, were convinced that it could not be verified. However, even if ratification were possible, reports suggest that even the compromise position, which would allow development of incapacitants for domestic riot control and law enforcement, was not totally secure:[23]

> . . . The CWC has caused some trepidation on the part of the US

chemical industry, which fears the pact will restrict the production of some commercial chemicals. In addition some in the Pentagon are also wary of some sections of the CWC that restrict the use of selected riot control agents.

The battle over this particular issue had clearly yet to be decided.

An idea of where we might end up, if incapacitant chemical (non-lethal) weapons get out of control, may be gained from a study of the use of non-lethal airpower for strategic paralysis mentioned earlier.[12] In a final section of their paper 'Future Challenges', the authors considered certain moral and legal issues. In their opinion:

> . . . current treaties must be renegotiated to take into account other non-lethal technologies. Certain chemical and biological uses of non-lethal technology may be acceptable . . .

Citing an expert on psychopharmacology, they suggest that the science of drugs which affect the mind is:

> . . . on 'the brink of revolution'. Previously, psychopharmacology had concentrated on the development of drugs that modify brain chemistry of mentally ill patients. . . . Presently, scientists are studying 'normal' brains . . .

Moreover, the new drugs for affecting normal brains:

> . . . are supposed to have no serious side effects and no addictive properties. Potentially, psychopharmacology has great application for nonlethal warfare and should be followed closely to ensure that its offensive and defensive potentials are well understood . . .

The authors concede that the United States may not choose to take this route, but add that others may nevertheless do so. They conclude their consideration of the issues with what appears to be a rhetorical question:

> In short, technologies of the future will be able to incapacitate humans. If this option is the most efficient way to obtain strategic objectives, should we limit its use?

It may be acknowledged that were there a safe, reliable way of dealing with some of the scenarios proposed for the employment of calmative agents, and no further potential repercussions to consider, there would be some merit in considering the option. Using a non-lethal alternative to handle a situation where combatants and non-combatants are mixed is, of course, more attractive than being able to do nothing or using potentially lethal force. That argument must be balanced against other considerations, particularly longer-term consequences, which could all too easily be ignored by those advocating the immediate benefits of deployment. We have to ask how the sentiments expressed in the question above will be perceived in other parts of the world. It will surely be thought that we wish to let networks of scientists[24] off the leash in advanced industrial countries so that they may develop whatever military applications the 'Decade of the Brain' makes possible. This would naturally be combined with a strategy of technology denial towards the rest of the world, but it is surely obvious that technology denial is impossible in this field of research. The technology is not that complex – particularly if one were not too concerned about just how 'non-lethal' one's agents would be in practice. The crucial point has been made by Michael Moodie, President of the Chemical and Biological Arms Control Institute, and former assistant director of the US Arms Control and Disarmament Agency. In his view:[25]

> . . . As the biotechnology revolution proceeds, the interest in developing countries plagued by endemic disease in combating serious global health problems will increase. As a result, biotechnology capabilities are likely to become more widespread, as will biological weapons capabilities, to which they are closely related . . .

Clearly, a number of countries could already turn their civil capabilities to the production of chemical and biological weapons if they so wished. The question is whether they will take the political decision to do so. In Moodie's opinion:

> . . . Ensuring that the incentives and disincentives of the international system are structured to channel those choices in stabilising and peaceful directions is a fundamental challenge of the post-cold war era.

Designing an appropriate framework is complex but, crucially, if we grasp the opportunity co-operatively to agree sensible international agreements and if we stick by them, we are likely to live in a more stable world. On the other hand, if we seize whatever military technical advantages we think available – particularly from the fields of modern neuroscience and biotechnology – we could indeed end up with a new form of warfare. Whether we would like what we had achieved is another matter.

This point was made with some force in a detailed study of non-lethal weapons by three US military officers during a year in residence at the Kennedy School of Government at Harvard University.[26] They argued that while the Chemical Weapons Convention contains a general prohibition on the use of riot control agents 'in international armed conflict as a method of warfare', and that use of such agents in a conflict requires presidential approval, 'it is the interpretation of the US that the Convention does not apply to peacekeeping or humanitarian operations'. Yet they also argued that most states targeted with non-lethal coercive measures may be unable to respond in kind and may resort to lethal weaponry or terrorism and thus trigger wider hostilities. In a chilling passage they concluded:

> . . . Of particular concern is the non-lethal use of biological or chemical agents (such as calmatives); a targeted state may feel justified in responding with weapons of mass destruction . . .

It bears repeating that the effects of chemical incapacitants on a large number of people might include some casualties which, to those attempting to assess events in a confused situation, could indicate that much more lethal material had been used. It should also be remembered, although we have not dwelt on it here, that there are many biological agents which could be used as incapacitants – indeed, some have been weaponised.[27] Resort to the use of such agents by an opponent would probably create even more difficulties in containing inadvertent escalation.

It is, of course, important to avoid wild speculation about the future possibilities that research might bring, but it is also necessary to consider carefully what might happen over the next decade or two. We have concentrated here on neuroscience and to a lesser extent on biotechnology, but the major scientific revolution in progress now – and the one the military is most interested in – is in information technology. It will not have escaped some readers' notice that there is a potential for interaction between all three. *Defense News*, for example, carried a story in early 1995 which began:[28]

> Battles of the future could be waged with genetically-engineered organisms . . . whose minds are controlled by computer chips engineered with living brain cells . . .

A recently retired associate director of the US Naval Research Laboratory was quoted as saying that at the laboratory:

> . . . research, called Hippocampal Neuron Patterning, grows live neurons on computer chips . . .

The story furthermore suggested that the battle scenario could be realised in 15 years if the Navy research paid off. That may sound far-fetched, but the Navy's short-term aim, of producing an 'electronic canary' which could go into contaminated areas and send out a signal from a chip when brain function altered in response to 'impending paralysis or death', may not be so far from reality. Such research on biosensors seems sensible and legitimate but it could be just the first step on a long road if misperceived abroad.

The overall conclusion then is that we must exercise great caution over non-lethal weapons options for forces of the advanced industrialised nations. Any means of reducing the casualties of conflict and war in an effective way deserves consideration, but many non-lethal options appear to be, or could be perceived to be, driven more by a desire to take military advantage of technical capabilities that these nations currently possess, and which others presently do not. Importantly, the development and fielding of many of these weapon systems will call into question the arms control regime we have so far managed to erect and will complicate efforts to proceed further with the necessary process of military de-escalation upon which stability and security will depend in the next century. Moreover, deployment of weapon systems originating from the pursuit of superiority in what are essentially areas of dual-use technology seem likely to generate responses in kind and will end in more suffering and not less.

Finally, it has to be said that while civilised countries may check the use of weapon systems carefully against the requirements of international law,[29] any weapons which advanced countries choose to develop will soon become widely available if the technology is not complex. Such weapons might then not be tightly regulated. CS gas, for example, was certainly used before the employment of lethal chemicals by Iraq against Iran in the 1980s[30] and, as we have seen in Chapter 5, it has been suggested that BZ may have been used to force the fall of Zepa[31] and Srebrenica[32] in Bosnia in 1995. The use of new non-lethal agents is not a distant, unthinkable danger.

APPENDIX I

THE INHUMANE WEAPONS CONVENTION

CONVENTION ON PROHIBITIONS OR RESTRICTIONS ON THE USE OF CERTAIN CONVENTIONAL WEAPONS WHICH MAY BE DEEMED TO BE EXCESSIVELY INJURIOUS OR TO HAVE INDISCRIMINATE EFFECTS AND PROTOCOLS (1980)

Entry into Force: 2 December 1983

The High Contracting Parties, Recalling that every State has the duty, in conformity with the Charter of the United Nations, to refrain in its international relations from the threat or use of force against the sovereignty, territorial integrity or political independence of any State, or in any other manner inconsistent with the purposes of the United Nations,

Further recalling the general principle of the protection of the civilian population against the effects of hostilities,

Basing themselves on the principle of international law that the right of the parties to an armed conflict to choose methods or means of warfare is not unlimited, and on the principle that prohibits the employment in armed conflicts of weapons, projectiles and material and methods of warfare of a nature to cause superfluous injury or unnecessary suffering,

Also recalling that it is prohibited to employ methods or means of warfare which are intended, or may be expected, to cause widespread, long-term and severe damage to the natural environment,

Confirming their determination that in cases not covered by this Convention and its annexed Protocols or by other international agreements, the civilian population and the combatants shall at all times remain under the protection and authority of the principles of international law derived from established custom, from the principles of humanity and from the dictates of public conscience,

Desiring to contribute to international detente, the ending of the arms race and the building of confidence among States, and hence to the realisation of the aspiration of all peoples to live in peace,

Recognizing the importance of pursuing every effort which may contribute to progress towards general and complete disarmament under strict and effective international control,

Reaffirming the need to continue the codification and progressive development of the rules of international law applicable in armed conflict,

Wishing to prohibit or restrict further the use of certain conventional weapons and believing that the positive results achieved in this area may facilitate the main talks on disarmament with a view to putting an end to the production, stockpiling and proliferation of such weapons,

Emphasizing the desirability that all States become parties to this Convention and its annexed Protocols, especially the militarily significant States,

Bearing in mind that the General Assembly of the United Nations and the United Nations Disarmament Commission may decide to examine the question of a possible broadening of the scope of the prohibitions and restrictions contained in this Convention and its annexed Protocols,

Further bearing in mind that the Committee on Disarmament may decide to consider the question of adopting further measures to prohibit or restrict the use of certain conventional weapons,

Have agreed as follows:

Article 1
Scope of application
This Convention and its annexed Protocols shall apply in the situations referred to in Article 2 common to the Geneva Conventions of 12 August 1949 for the Protection of War Victims, including any situation described in paragraph 4 of Article 1 of Additional Protocol I to these Conventions.

Article 2
Relations with other international agreements
Nothing in this Convention or its annexed Protocols shall be interpreted as detracting from other obligations imposed upon the High Contracting Parties by international humanitarian law applicable in armed conflict.

Article 3
Signature
This Convention shall be open for signature by all States at United Nations Headquarters in New York for a period of twelve months from 10 April 1981.

Article 4
Ratification, acceptance, approval or accession
1. This Convention is subject to ratification, acceptance or approval by the Signatories. Any State which has not signed this Convention may accede to it.
2. The instruments of ratification, acceptance, approval or accession shall be deposited with the Depositary.
3. Expressions of consent to be bound by any of the Protocols annexed to this Convention shall be optional for each State, provided that at the time of the deposit of its instrument of ratification, acceptance or approval of this Convention or of accession thereto, that State shall notify the Depositary of its consent to be bound by any two or more of these Protocols.
4. At any time after the deposit of its instrument of ratification, acceptance or approval of this Convention or of accession thereto, a State may notify the Depositary of its consent to be bound by any annexed Protocol by which it is not already bound.
5. Any Protocol by which a High Contracting Party is bound shall for that Party form an integral part of this Convention.

Article 5
Entry into force
1. This Convention shall enter into force six months after the date of deposit of the twentieth instrument of ratification, acceptance, approval or accession.
2. For any State which deposits its instrument of ratification, acceptance, approval or accession after the date of the deposit of the twentieth instrument of ratification, acceptance, approval or accession, this Convention shall enter into force six months after the date on which that State has deposited its instrument of ratification, acceptance, approval or accession.
3. Each of the Protocols annexed to this Convention shall enter into force six months after the date by which twenty States have notified their consent to be bound by it in accordance with paragraph 3 or 4 of Article 4 of this Convention.

4. For any State which notifies its consent to be bound by a Protocol, annexed to this Convention after the date by which twenty States have notified their consent to be bound by it, the Protocol shall enter into force six months after the date on which that State has notified its consent so to be bound.

Article 6
Dissemination

The High Contracting Parties undertake, in time of peace as in time of armed conflict, to disseminate this Convention and those of its annexed Protocols by which they are bound as widely as possible in their respective countries and, in particular, to include the study thereof in their programmes of military instruction, so that those instruments may become known to their armed forces.

Article 7
Treaty relations upon entry into force of this Convention

1. When one of the parties to a conflict is not bound by an annexed Protocol, the parties bound by this Convention and that annexed Protocol shall remain bound by them in their mutual relations.
2. Any High Contracting Party shall be bound by this Convention and any Protocol annexed thereto which is in force for it, in any situation contemplated by Article 1, in relation to any State which is not a party to this Convention or bound by the relevant annexed Protocol, if the latter accepts and applies this Convention or the relevant Protocol, and so notifies the Depositary.
3. The Depositary shall immediately inform the High Contracting Parties concerned of any notification received under paragraph 2 of this Article.
4. This Convention, and the annexed Protocols by which a High Contracting Party is bound, shall apply with respect to an armed conflict against that High Contracting Party of the type referred to in Article 1, paragraph 4, of Additional Protocol I to the Geneva Conventions of 12 August 1949 for the Protection of War Victims:
    (a) where the High Contracting Party is also a party to Additional Protocol I and an authority referred to in Article 96, paragraph 3, of that Protocol has undertaken to apply the Geneva Conventions and Additional Protocol I in accordance with Article 96, paragraph 3, of the said Protocol, and undertakes to apply this Convention and the relevant annexed Protocols in relation to that conflict; or

(b) where the High Contracting Party is not a party to Additional Protocol I and an authority of the type referred to in subparagraph (a) above accepts and applies the obligations of the Geneva Conventions and of this Convention and the relevant annexed Protocols in relation to that conflict. Such an acceptance and application shall have in relation to that conflict the following effects:

(i) the Geneva Conventions and this Convention and its relevant annexed Protocols are brought into force for the parties to the conflict with immediate effect;

(ii) the said authority assumes the same rights and obligations as those which have been assumed by a High Contracting Party to the Geneva Conventions, this Convention and its relevant annexed Protocols; and

(iii) the Geneva Conventions, this Convention and its relevant annexed Protocols are equally binding upon all parties to the conflict.

The High Contracting Party and the authority may also agree to accept and apply the obligations of Additional Protocol I to the Geneva Conventions on a reciprocal basis.

Article 8
Review and amendments

1. (a) At any time after the entry into force of this Convention any High Contracting Party may propose amendments to this Convention or any annexed Protocol by which it is bound. Any proposal for an amendment shall be communicated to the Depositary, who shall notify it to all the High Contracting Parties and shall see their views on whether a conference should be convened to consider the proposal. If a majority, that shall not be less than eighteen of the High Contracting Parties so agree, he shall promptly convene a conference to which all High Contracting Parties shall be invited. States not parties to this Convention shall be invited to the conference as observers.

(b) Such a conference may agree upon amendments which shall be adopted and shall enter into force in the same manner as this Convention and the annexed Protocols, provided that amendments to this Convention may be adopted only by the High Contracting Parties and that amendments to a specific annexed Protocol may be adopted only by the High Contracting Parties which are bound by that Protocol.

2.   (a) At any time after the entry into force of this Convention any High Contracting Party may propose additional protocols relating to other categories of conventional weapons not covered by the existing annexed Protocols. Any such proposal for an additional protocol shall be communicated to the Depositary, who shall notify it to all the High Contracting Parties in accordance with subparagraph 1 (a) of this Article. If a majority, that shall not be less than eighteen of the High Contracting Parties so agree, the Depositary shall promptly convene a conference to which all States shall be invited.

  (b) Such a conference may agree, with the full participation of all States represented at the conference, upon additional protocols which shall be adopted in the same manner as this Convention, shall be annexed thereto and shall enter into force as provided in paragraphs 3 and 4 of Article 5 of this Convention.

3.   (a) If, after a period of ten years following the entry into force of this Convention, no conference has been convened in accordance with subparagraph 1 (a) or 2 (a) of this Article, any High Contracting Party may request the Depositary to convene a conference to which all High Contracting Parties shall be invited to review the scope and operation of this Convention and the Protocols annexed thereto and to consider any proposal for amendments of this Convention or of the existing Protocols. States not Parties to this Convention shall be invited as observers to the conference. The conference may agree upon amendments which shall be adopted and enter into force in accordance with subparagraph 1 (b) above.

  (b) At such conference consideration may also be given to any proposal for additional protocols relating to other categories of conventional weapons not covered by the existing annexed Protocols. All States represented at the conference may participate fully in such consideration. Any additional protocols shall be adopted in the same manner as this Convention, shall be annexed thereto and shall enter into force as provided in paragraphs 3 and 4 of Article 5 of this Convention.

  (c) Such a conference may consider whether provision should be made for the convening of a further conference at the request of any High Contracting Party if, after a similar period to that referred to in subparagraph 3 (a) of this Article, no conference has been convened in accordance with subparagraph 1 (a) or 2 (a) of this Article.

Article 9
Denunciation

1.   Any High Contracting Party may denounce this Convention or any of its annexed Protocols by so notifying the Depositary.
2.   Any such denunciation shall only take effect one year after receipt by the Depositary of the notification of denunciation. If, however, on the expiry of that year the denouncing High Contracting Party is engaged in one of the situations referred to in Article 1, the Party shall continue to be bound by the obligations of this Convention and of the relevant annexed Protocols until the end of the armed conflict or occupation and, in any case, until the termination of operations connected with the final release, repatriation or re-establishment of the person protected by the rules of international law applicable in armed conflict, and in the case of any annexed Protocol containing provisions concerning situations in which peace-keeping, observation or similar functions are performed by United Nations forces or missions in the area concerned, until the termination of those functions.
3.   Any denunciation of this Convention shall be considered as also applying to all annexed Protocols by which the denouncing High Contracting Party is bound.
4.   Any denunciation shall have effect only in respect of the denouncing High Contracting Party.
5.   Any denunciation shall not affect the obligations already incurred, by reason of an armed conflict, under this Convention and its annexed Protocols by such denouncing High Contracting Party in respect of any act committed before this denunciation becomes effective.

Article 10
Depositary

1.   The Secretary-General of the United Nations shall be the Depositary of this Convention and of its annexed Protocols.
2.   In addition to his usual functions, the Depositary shall inform all States of:
     (a) signatures affixed to this Convention under Article 3;
     (b) deposits of instruments of ratification, acceptance or approval of or accession to this Convention deposited under Article 4;
     (c) notifications of consent to be bound by annexed Protocols under Article 4;
     (d) the dates of entry into force of this Convention and of each of its annexed Protocols under Article 5; and

(e) notifications of denunciation received under Article 9, and their effective date.

Article 11
Authentic texts

The original of this Convention with the annexed Protocols, of which the Arabic, Chinese, English, French, Russian and Spanish texts are equally authentic, shall be deposited with the Depositary, who shall transmit certified true copies thereof to all States.

PROTOCOL ON NON-DETECTABLE FRAGMENTS (PROTOCOL I)

It is prohibited to use any weapon the primary effect of which is to injure by fragments which in the human body escape detection by X-rays.

PROTOCOL ON PROHIBITIONS OR RESTRICTIONS ON THE USE OF MINES, BOOBY TRAPS AND OTHER DEVICES (PROTOCOL II)

Article 1
Material scope of application

This Protocol relates to the use on land of the mines, booby-traps and other devices defined herein, including mines laid to interdict beaches, waterway crossings or river crossings, but does not apply to the use of anti-ship mines at sea or in inland waterways.

Article 2
Definitions
For the purpose of this Protocol:

1. 'Mine' means any munition placed under, on or near the ground or other surface area and designed to be detonated or exploded by the presence, proximity or contact of a person or vehicle, and 'remotely delivered mine' means any mine so defined delivered by artillery, rocket, mortar or similar means or dropped from an aircraft.

2. 'Booby-trap' means any device or material which is designed, constructed or adapted to kill or injure and which functions unexpectedly when a person disturbs or approaches an apparently harmless object or performs an apparently safe act.

3. 'Other devices' means manually-emplaced munitions and devices designed to kill, injure or damage and which are actuated by remote control or automatically after a lapse of time.

4. 'Military objective' means, so far as objects are concerned, any object which by its nature, location, purpose or use makes an effective contribution to military action and whose total or partial destruction, capture or neutralisation, in the circumstances ruling at the time, offers a definite military advantage.
5. 'Civilian objects' are all objects which are not military objectives as defined in paragraph 4.
6. 'Recording' means a physical, administrative and technical operation designed to obtain, for the purpose of registration in the official records, all available information facilitating the location of minefields, mines and booby-traps.

Article 3
General restrictions on the use of mines, booby-traps and other devices

1. This Article applies to:
    (a) mines;
    (b) booby-traps; and
    (c) other devices.
2. It is prohibited in all circumstances to direct weapons to which this Article applies, either in offence, defence or by way of reprisals, against the civilian population as such or against individual civilians.
3. The indiscriminate use of weapons to which this Article applies is prohibited. Indiscriminate use is any placement of such weapons:
    (a) which is not on, or directed at, a military objective; or
    (b) which employs a method or means of delivery which cannot be directed at a specific military objective; or
    (c) which may be expected to cause incidental loss of civilian life, injury to civilians, damage to civilian objects, or a combination thereof, which would be excessive in relation to the concrete and direct military advantage anticipated.
4. All feasible precautions shall be taken to protect civilians from the effects of weapons to which this Article applies. Feasible precautions are those precautions which are practicable or practically possible taking into account all circumstances ruling at the time, including humanitarian and military considerations.

Article 4
Restrictions on the use of mines other than remotely delivered mines, booby-traps and other devices in populated areas
1. This Article applies to:
    (a) mines other than remotely delivered mines;
    (b) booby-traps; and
    (c) other devices.
2. It is prohibited to use weapons to which this Article applies in any city, town, village or other area containing a similar concentration of civilians in which combat between ground forces is not taking place or does not appear to be imminent, unless either:
    (a) they are placed on or in the close vicinity of a military objective belonging to or under the control of an adverse party; or
    (b) measures are taken to protect civilians from their effects, for example, the posting of warning signs, the posting of sentries, the issue of warnings or the provision of fences.

Article 5
Restrictions on the use of remotely delivered mines
1. The use of remotely delivered mines is prohibited unless such mines are only used within an area which is itself a military objective or which contains military objectives, and unless:
(a) their location can be accurately recorded in accordance with Article 7(1)(a); or
(b) an effective neutralizing mechanism is used on each such mine, that is to say, a self-actuating mechanism which is designed to render a mine harmless or cause it to destroy itself when it is anticipated that the mine will no longer serve the military purpose for which it was placed in position, or a remotely-controlled mechanism which is designed to render harmless or destroy a mine when the mine no longer serves the military purpose for which it was placed in position.
2. Effective advance warning shall be given of any delivery or dropping of remotely delivered mines which may affect the civilian population, unless circumstances do not permit.

Article 6
Prohibition on the use of certain booby traps
1. Without prejudice to the rules of international law applicable in armed conflict relating to treachery and perfidy, it is prohibited in all circumstances to use:

(a) any booby-trap in the form of an apparently harmless portable object which is specifically designed and constructed to contain explosive material and to detonate when it is disturbed or approached; or

(b) booby-traps which are in any way attached to or associated with:
    (i) internationally recognised protective emblems, signs or signals;
    (ii) sick, wounded or dead persons;
    (iii) burial or cremation sites or graves;
    (iv) medical facilities, medical equipment, medical supplies or medical transportation;
    (v) children's toys or other portable objects or products specially designed for the feeding, health, hygiene, clothing or education of children;
    (vi) food or drink;
    (vii) kitchen utensils or appliances except in military establishments, military locations or military supply depots;
    (viii) objects clearly of a religious nature;
    (ix) historic monuments, works of art or places or worship which constitute the cultural or spiritual heritage of peoples;
    (x) animals or their carcasses.

2. It is prohibited in all circumstances to use any booby-trap which is designed to cause superfluous injury or unnecessary suffering.

Article 7
Recording and publication of the location of minefields, mines and booby-traps

1. The parties to a conflict shall record the location of:
    (a) all pre-planned minefields laid by them; and
    (b) all areas in which they have made large-scale and pre-planned use of booby-traps.

2. The parties shall endeavour to ensure the recording of the location of all other minefields, mines and booby-traps which they have laid or placed in position.

3. All such records shall be retained by the parties who shall:
    (a) immediately after the cessation of active hostilities:
        (i) take all necessary and appropriate measures, including the use of such records, to protect civilians from the effects of minefields, mines and booby-traps; and either
        (ii) in cases where the forces of neither party are in the territory of the adverse party, make available to each

              other and to the Secretary-General of the United Nations all information in their possession concerning the location of minefields, mines and booby-traps in the territory of the adverse party; or

    (iii) once complete withdrawal of the forces of the parties from the territory of the adverse party has taken place, make available to the adverse party and to the Secretary-General of the United Nations all information in their possession concerning the location of minefields, mines and booby traps in the territory of the adverse party;

(b) When a United Nations force or mission performs functions in any area, make available to the authority mentioned in Article 8 such information as is required by that Article;

(c) whenever possible, by mutual agreement provide for the release of information concerning the location of minefields, mines and booby traps, particularly in agreements governing the cessation of hostilities.

Article 8
Protection of United Nations forces and missions from the effects of minefields, mines and booby traps

1. When a United Nations force or mission performs functions of peacekeeping, observation or similar functions in any area, each party to the conflict shall if requested by the head of the United Nations force or mission in that area, as far as it is able:

    (a) remove or render harmless all mines or booby traps in that area;

    (b) take such measures as may be necessary to protect the force or mission from the effects of minefields, mines and booby traps while carrying out its duties; and

    (c) make available to the head of the United Nations force or mission in that area, all information in the party's possession concerning the location of minefields, mines and booby traps in that area.

2. When a United Nations fact-finding mission performs functions in any area, any party to the conflict concerned shall provide protection to that mission except where, because of the size of such mission, it cannot adequately provide such protection. In that case it shall make available to the head of the mission the information in its possession concerning the location of minefields, mines and booby-traps in that area.

Article 9
International co-operation in the removal of minefields, mines and booby traps
After the cessation of active hostilities, the parties shall endeavour to reach agreement, both among themselves and, where appropriate, with other States and with international organisations, on the provision of information and technical and material assistance – including, in appropriate circumstances, joint operations necessary to remove or otherwise render ineffective minefields, mines and booby-traps placed in position during the conflict.

TECHNICAL ANNEX TO THE PROTOCOL ON PROHIBITIONS OR RESTRICTIONS ON THE USE OF MINES, BOOBY TRAPS AND OTHER DEVICES (PROTOCOL II)

Whenever an obligation for the recording of the location of minefields, mines and booby traps arises under the Protocol, the following guidelines shall be taken into account.
1. With regard to pre-planned minefields and large-scale and pre-planned use of booby traps:
    (a) maps, diagrams or other records should be made in such a way as to indicate the extent of the minefield or booby-trapped area; and
    (b) the location of the minefield or booby-trapped area should be specified by relation to the co-ordinates of a single reference point and by the estimated dimensions of the area containing mines and booby traps in relation to that single reference point.
2. With regard to other minefields, mines and booby traps laid or placed in position:
In so far as possible, the relevant information specified in paragraph 1 above should be recorded so as to enable the areas containing minefields, mines and booby traps to be identified.

PROTOCOL ON PROHIBITIONS OR RESTRICTIONS ON THE USE OF INCENDIARY WEAPONS (PROTOCOL III)

Article 1
Definitions
For the purpose of this Protocol:
1. 'Incendiary weapon' means any weapon or munition which is primarily designed to set fire to objects or to cause burn injury to

persons through the action of flame, heat, or a combination thereof, produced by a chemical reaction of a substance delivered on the target.

(a) Incendiary weapons can take the form of, for example, flame throwers, fougasses, shells, rockets, grenades, mines, bombs and other containers of incendiary substances.

(b) Incendiary weapons do not include:

    (i) Munitions which may have incidental incendiary effects, such as illuminants, tracers, smoke or signalling systems;

    (ii) Munitions designed to combine penetration, blast or fragmentation effects with an additional incendiary effect, such as armour-piercing projectiles, fragmentation shells, explosive bombs and similar combined-effects munitions in which the incendiary effect is not specifically designed to cause burn injury to persons, but to be used against military objectives, such as armoured vehicles, aircraft and installations or facilities.

2. 'Concentration of civilians' means any concentration of civilians, be it permanent or temporary, such as in inhabited parts of cities, or inhabited towns or villages, or as in camps or columns of refugees or evacuees, or groups of nomads.

3. 'Military objective' means, so far as objects are concerned, any object which by its nature, location, purpose or use makes an effective contribution to military action and whose total or partial destruction capture or neutralisation, in the circumstances ruling at the time, offers a definite military advantage.

4. 'Civilian objects' are all objects which are not military objectives as defined in paragraph 3.

5. 'Feasible precautions' are those precautions which are practicable or practically possible taking into account all circumstances ruling at the time, including humanitarian and military considerations.

Article 2
Protection of civilians and civilian objects

1. It is prohibited in all circumstances to make the civilian population as such, individual civilians or civilian objects the object of attack by incendiary weapons.

2. It is prohibited in all circumstances to make any military objective located within a concentration of civilians the object of attack by air-delivered incendiary weapons.

3. It is further prohibited to make any military objective located within

a concentration of civilians the object of attack by means of incendiary weapons other than air-delivered incendiary weapons, except when such military objective is clearly separated from the concentration of civilians and all feasible precautions are taken with a view to limiting the incendiary effects to the military objective and to avoiding, and in any event to minimizing, incidental loss of civilian life, injury to civilians and damage to civilian objects.

4. It is prohibited to make forests or other kinds of plant cover the object of attack by incendiary weapons except when such natural elements are used to cover, conceal or camouflage combatants or other military objectives, or are themselves military objectives.

APPENDIX II

CHEMISTRY OF PSYCHOCHEMICAL AGENTS*

A SUMMARY OF THE CHEMICAL AND RADIOLOGICAL LABORATORIES' PROGRESS FOR THE FISCAL YEARS 1950 THROUGH 1955

1. Introduction

The possible military usefulness of affecting the minds of masses of troops by the use of chemical agents was developed in 1949 in a preliminary report by Dr L.W.Greene, Technical Director, Chemical and Radiological Laboratories. Applications of the concept were outlined, and a preliminary discussion of mental abnormalities of interest, as well as means by which they might be produced, was presented. During the period from FY1950 through FY1955, feasibility studies in this new field were conducted. Literature studies were made, plans for the chemical approach to the problem were formulated and some experimental work on the preparation (by our Laboratories) and testing (by Medical Laboratories) of candidate compounds was accomplished. A summary of our work during this period is presented in the following sections.

2. Literature Studies and Chemical Investigations

In order to give a firmer background for further literature search and experimental work and to clarify and crystallize our ideas as to desirable properties for psychochemical agents, a study of the symptoms observed

* From note 44 of Chapter 5

in various well-known abnormal mental states was undertaken. One of the results of this study was an elaboration of Dr Greene's original list of abnormal symptoms of potential military significance. This served as a guide in the simultaneously proceeding search of the chemical literature and as a basis for the following list of useful symptoms furnished later by an Ad Hoc Subcommittee of the Advisory Committee for New Agents.

(a) Delusions, hallucinations, mania, delirium
(b) Defects in hearing, sight, and/or judgment
(c) Psychosis, depression, suicidal tendencies
(d) Inco-ordination, paralysis (spastic, flaccid)
(e) Convulsions
(f) Anemia, listlessness, weakness
(g) Headache, nausea, dizziness
(h) Cutaneous disorders, such as urticaria

Our literature studies started with a survey of mescaline, 3,4,5-trimethoxyphenethyl amine and related substances. This work was undertaken to become better acquainted with the properties of a typical psychically active compound and to form the basis for future experimental work. The investigation revealed that mescaline produces its interesting and well-known hallucinatory and other effects in a dose range of 200 to 500 mg per man. Doses in this range are normally considered to be too high for chemical warfare purposes. However, on the basis of this activity and that of other phenethylamines, it was proposed as a practical research problem to study alterations of the mescaline structure in the hope of finding a derivation of sufficient potency to be of definite interest.

Shortly after the mescaline work was started (fall of 1949), our attention was directed to lysergic acid diethylamide (LSD) as a psychically active substance of low toxicity which is effective at the very low dose range. . . . Surveys were made of this substance and compounds related to it and of other active compounds, both naturally occurring and synthetic, described in the chemical and pharmacological literature. The work suggested that, although there are a vast numbers of compounds which may produce mental disturbances, the most favorable areas for our preliminary work could probably be restricted to a few. The following three fields were indicated as being most likely to yield with minimum delay coumpounds suitable for investigation as possible candidate psychochemical agents:

(a) *Indole derivatives* related to LSD and harmine,
(b) *Benzopyran derivatives* related to tetrahydrocannabinol,
(c) *Phenethyl amines* related to mescaline, adrenalin, etc.

Our chemical approach to the problem thus consisted of literature study on appropriate members of selected classes of compounds, the synthesis or other procurement of substances in accordance with the planned program and the submission of our materials to Medical Laboratories, with which our plans are closely co-ordinated, for evaluation.

As of today, lysergic acid diethylamide is by far the most potent psychochemical yet discovered, and no known synthetic of comparatively simple structure competes with it. Because of the properties of LSD and other psychically active substances, e.g., harmine and yohimbine, which are also members of the group of indole derivatives, it is believed that this field is the most promising one for exploitation.

Investigation of the literature also showed that in the study of benzopyran derivatives related to tetrahydrocannabinol, the active principle of marihuana, several homologs were prepared which had considerably higher activity. Tests carried out on one of these, a C9 homolog, showed it to have a potency which we believe makes this group second only to the indole series in immediate interest and importance to the problem.

The phenethyl amine series appears now to be of much less interest as a practical source of candidate substances than the other two, because of the much larger dose required to produce desired effects. However, two recent suggestions have appeared which show that these derivatives may yet prove to be of considerable importance in our problem. These are (a) that some abnormality in the function of the adrenalin system may be the key factor in mental illness and that the adrenalin metabolites, adrenochrome and/or adrenoxine (or derivatives of these) may be the causative agents and (b) that serotonin (5-hydroxytryptamine) may play a vital role in brain function, and that LSD, harmine, yohimbine, etc. exert their mentally deranging effect by competitively inhibiting serotonin. Since adrenochrome is an indole derivative, it may be that the phenethyl amine derivatives are more closely related to the indoles biochemically than first appeared from their structures.

On the experimental side, the problem has been handled as a feasibility study, and our volume of work has accordingly not been high. It is summarised briefly as follows.

(a) *Phenethyl Amine Derivatives.* A total of 25 to 30 methoxy- and methlenedioxyphenethylamines were prepared and forwarded to Medical Laboratories for testing. The list includes:
   1. Mono-, di-, and trimethoxy and methylenedioxy-phenethyl amines.
   2. Mono-, di-, and trimethoxy- and methylenedioxy-phenyl isopropyl amines.

    3. Mono-, di-, and trimethoxy- and methyleneddioxy butyl amines.

    4. A few N-Methyl derivatives of these substances.

The pharmacological properties of these compounds are discussed in Appendix IV. It can be stated here that none of these tested proved to be sufficiently active to make it of interest for CW purposes, and it is improbable that extensive *unguided* investigations in this group will be profitable.

(b) *Benzopyran Derivatives*. Only five or six tetrahydrocannabinol homologs were prepared under contract. The list included tetrahydrocannabinol itself and the highly active C9 (1,2-dimethyl heptyl) homolog mentioned above.

(c) *Indole Derivatives*. LSD has been made available, and a few simpler compounds corresponding to or related to fragments of the LSD structrure were synthesized at Chemical and Radiological Laboratories and under contract. Also, a number of LSD intermediates, fragments and derivatives were submitted by an interested outside source. A total of perhaps two dozen such compounds have been supplied. LSD itself has been extensively tested under Chemical Corps sponsorship and in many centers of psychiatric research, and, as indicated above, it is considered that from the point of view of its effects in extremely low doses it would be a valuable psychochemical agent. It must be borne in mind that this substance and its salts are high-melting solids and therefore present dissemination problems which may be difficult to solve; however, the effectiveness of LSD when inhaled as a sprayed mist has been demonstrated.

Finally, although LSD has been synthesized and a natural growth method of production is available, it may present very difficult, if perhaps not insurmountable, problems of availability. For these reasons, a highly active compound, easily prepared and of practical physical and chemical properties, should be intensively sought. Biological results on two simple alkyl derivatives of indole ethyl amine showed these substances to have considerable mescalinelike activity in animals, thus confirming our belief that investigations in the indole field are of the most interest to this problem. Tests of other members of this group are now in progress.

As indicated above, this problem has been handled up to the present date as a feasibility study. With the approval of a new subproject, 4-08-03-016-05, Psychochemical Agents, on 30 June 1955, it became an official part of the agents program. The following military characteristics are set forth in the approved project data sheet.

1. *Type of Action:* Psychochemical agents should cause temporary mental and/or motor incapacitation (such as a state of suspended animation) of personnel.

2. *Rapidity of Action:* The agents should act promptly after administration, preferably in less than an hour. For sabotage (covert) operations, agents with delayed action (several hours) prior to onset of symptoms may be useful.

3. *Duration of Action:* It is desirable, but not essential, that the agents have no permanent effects. The minimum time for duration of action (symptoms) should be 24 hours. Agents whose action lasts for several days (even a week) may be desirable for some purposes.

4. *Effectiveness:* The agents should have a potency at least equal to the nerve gases. This means that a dosage of 0.50 to 0.75 mg should incapacitate a man. This should not restrict research and experimentation on less potent compounds which may furnish data and information leading to greater understanding of and/or better psychochemicals.

5. *Toxicity:* The agents should have a low intrinsic toxicity; however, comparatively high toxicity should not be the basis for discarding substances having the desired incapacitating properties.

6. *Dissemination:* The agents should be capable of being disseminated in airborne form under all environmental conditions likely to be encountered.

7. *Stability:* The agents should be capable of being stored for long periods of time under all extremes of environmental conditions.

3. Future Plans and Other Areas of Interest

As stated previously, but little exploratory work has been accomplished among indole derivatives, except for the extensive investigations on LSD, and among tetrahydrocannabinol homologs and analogs. It is considered that LSD and the C9 benzopyran derivative (relative of tetrahydrocannabinol) should be fully exploited and that more extensive synthetic programs based on available leads should be undertaken in the fields of indole and benzopyran derivatives.

In addition to the foregoing fields, the following are of interest, and it is believed their value to the problem should be assessed:

    (a) Sublethal effects of G-type agents

    (b) Curare and related synthetic paralysants

    (c) Thyroid and antithyroid substances; other hormones

    (d) Intoxicating principle of mushrooms, especially Amanita Muscaria

    (e) Substances related to bulbocapaine

    (f) Substances related to Belladonna alkaloids

The approach outlined above is seen to be largely a direct empirical one based on literature leads and a minimum of experimental work.

Psychiatrists, pharmacologists and biochemists are actively investigating the problem of the mechanism of mental illness, and as these studies progress leads to more effective psychochemical agents will undoubtedly become available. As this occurs, our program will be modified accordingly. The pharmacological and clinical studies certainly point in the direction of better understanding of these matters, and we therefore look forward with a good deal of optimism regarding the possibility of fulfilling military requirements in this field.

REFERENCES

Preface

1. Scott, W. B. (1995), 'Panel's report backs nonlethal weapons', *Aviation Week and Space Technology*, **143**, 16 October, pp. 60–2.
2. Nakamura, H. and Dando, M. R. (1993), 'Japan's military research and development: A high technology deterrent', *Pacific Review*, **6** (2), pp. 177–90.
3. Ricks, T. E. (1993), 'Nonlethal arms: New class of weapons could incapacitate foe yet limit casualties; military sees role for lasers, electromagnetic pulses, other high-tech tricks, sticky roads, stalled tanks', *Wall Street Journal*, 4 January, Section A, p. 1.
4. International Committee of the Red Cross (1994), *Report of the ICRC for the Review Conference of the 1980 United Nations Convention on Prohibitions or Restrictions on the Use of Certain Conventional Weapons which may be deemed to be Excessively Injurious or to have Indiscriminate Effects*, ICRC, Geneva.
5. United Nations (1995), *Round-up of Conference: Review Conference of States Parties to Conventional Weapons Convention, Vienna, 25 September–13 October*, DC/2535, 17 October.
6. Oxburgh, E. R. (1993) 'Future military technology and the West', *Royal United Services Institute Journal*, December, pp. 49–55.
7. Kellman, B. (1994), 'Bridling the international trade of catastrophic weaponry', *American University Law Review*, **43**, pp. 755–847.

Chapter 1: Peacekeeping in Chaotic Conflicts

1. Bowcott, O. (1995), 'Serb TV shows British captives', *The Guardian*, 30 May, p. 1.
2. Young, H. (1995), 'Sabres rattle on in a world out of control', *The Guardian*, 1 June, p. 15.

3. Bright, M. (1995), 'Struggling to keep the peace', *The Guardian* (Education), 10 January, p. 10.

4. Moore, M. (1995), 'UN peacekeeping: A glass half empty, half full', *Bulletin of the Atomic Scientists*, March/April, pp. 22–3.

5. Woodhouse, T. (1995), 'Peacekeeping and humanitarian intervention', paper presented to the conference 'Australia, the United Nations, Peacekeeping and Peacemaking', held at the University of Western Australia and Curtin University of Technology, May.

6. Anon. (1995), 'United Nations peacekeeping operations: History, resources, missions, and components', *International Defense Review – Defense '95*, pp. 119–27.

7. Roberts, A. (1994), 'The crisis in UN peacekeeping', *Survival*, **36** (3), pp. 93–120.

8. *United Nations (1982), 'Charter of the United Nations', Yearbook of the United Nations*, United Nations, New York.

9. Whitman, J. and Bartholomew, I. (1994), 'Collective control of UN peace support operations: A policy proposal', *Security Dialogue*, **25** (1), pp. 77–92.

10. James, A. (1993), 'The history of peacekeeping: An analytic perspective', in the *1993 Peacekeeping Symposium: Peacekeeping Norms, Policy and Process*, pp. 23–36, Centre for International and Strategic Studies, York University, Canada.

11. Helman, G. B. and Ratner, S. R. (1993), 'Saving failed states', *Foreign Policy*, **89**, pp. 3–20.

12. Maynes, C. W. (1993), 'Containing ethnic conflict', *Foreign Policy*, **90**, pp. 3–21.

13. Cooper, R. and Berdal, M. (1993), 'Outside intervention in ethnic conflicts', *Survival*, **35** (1), pp. 118–42.

14. Traynor, I. (1995), 'Rampant Serbs push UN aside', *The Guardian*, 12 July, p. 1.

15. Finnemore, M. (1995), 'Changing patterns of military intervention', paper presented at the 36th Annual Convention of the International Studies Association, Chicago, February.

16. Mandelbaum, M. (1994), 'The reluctance to intervene', *Foreign Policy*, **95**, pp. 3–18.

17. Rogers, P. and Dando, M. R. (1992), *A Violent Peace: Global Security After the Cold War*, Brassey's, London.

18. Kull, S. (1995), 'Misreading the public mood', *Bulletin of the Atomic Scientists*, March/April, pp. 55–9.

19. Policy Debates and Issues (1995), 'Agenda for Peace Supplement', *International Peacekeeping News*, January, pp. 16–17.

20. Schnabel, A. (1995), 'The emperor's new clothes? The United Nations and

• REFERENCES •

post-Cold War peacekeeping', paper presented at the 36th Annual Convention of the International Studies Association, Chicago, February.

Chapter 2: Benign Interventions with Non-lethal Weapons?

1. Barry, J. and Morganthau, T. (1994), 'Soon, "Phasers on stun"', *Newsweek*, 7 February, pp. 26–8.

2. Anon. (1994), 'Secret weapons for the CNN era', *Harper's Magazine*, October, p. 17.

3. Tapscott, M. and Atwal, K. (1993), 'New weapons that win without killing on DoD's horizon', *Defense Electronics*, February, pp. 41–6.

4. Lt. Col. Celick, A. J. (1968), 'Humane warfare for international peacekeeping', *Air University Review*, Sept./Oct., pp. 91–3.

5. Nunn, A. C. (1965), 'The arming of an international police', *Journal of Peace Research*, **2**, pp. 187–91.

6. Alexander, J. B. (1993), *Nonlethal Weapons and Limited Force Options*, LA–UR 93–3747, Los Alamos National Laboratory.

7. Anon. (1994), 'ARL technology handbook provides inside glimpse of army's non lethal weapons program direction', *Tactical Technology*, 22 June.

8. Anon. (1993), 'Non-lethal technologies', *Tactical Technology*, 3 February.

9. Evancoe, P. R. (1993), 'Non-lethal technologies enhance warrior's punch', *National Defense*, December, pp. 26–9.

10. Kiernan, V. (1993), 'War over weapons that can't kill', *New Scientist*, 11 December, pp. 14–16.

11. Wright, J. (1994), 'Shoot not to kill', *The Guardian* (Online), 19 May, p. 5.

12. Starr, B. (1994), 'Pentagon maps non-lethal options', *International Defense Review*, **7**, pp. 30–2.

13. Knoth, A. (1994), 'Disabling technologies: a critical assessment', *International Defense Review*, **7**, pp. 33–8.

14. Morris, J. and Morris, C. (1994), *Nonlethality: a global strategy*, Mimeo.

15. Kokoski, R. (1994), 'Non-lethal weapons: a case study of new technology developments', *SIPRI Yearbook*, pp. 367–86.

16. Aftergood, S. (1994), 'The soft-kill fallacy', *Bulletin of the Atomic Scientists*, September/October, pp. 40–5.

17. Shukman, D. (1995), *The Sorcerer's Challenge*, Hodder & Stoughton, London.

18. Anon. (1995), 'Blinding laser weapons: The need to ban a cruel and inhumane weapon', *Human Rights Watch Arms Project*, **7** (1), pp. 1–54.

19. Fulghum, D. A. (1993), 'EMP weapons lead race for non–lethal technology', *Aviation Week and Space Technology*, 24 May, p. 61.

20. Morris, J., Krivorotov, V. and Morris, C. (1993), *The Age of Chaos: Threat and Solution Beyond Containment*, Mimeo.

21. Tansey, G., Tansey, K. and Rogers, P. (1994), *A World Divided: Militarism and Development after the Cold War*, Earthscan, London.

22. Morrison, D. C. (1992), 'Focus: National Security. War without death', *National Journal*, 7 November.

23. US Army (1991), *Military Operations: Operations Concept for Non-Lethal Capabilities*, TRADOC Pamphlet 525–XX, Fort Monroe, Virginia.

24. Anon. (1994), 'DoD/DoJ accord may ease LEA access to advanced technology', *Tactical Technology*, 30 March.

25. Fulghum, D. A. (1992), 'U.S. weighs use of nonlethal weapons in Serbia if U.N. decides to fight', *Aviation Week and Space Technology*, 17 August, p. 62.

26. Kemey, G. and Dugan, M. J. (1992), 'Operation Balkan Storm: Here's a plan', *New York Times*, 29 November, p. 11.

27. Oxburgh, E. R. (1992), 'Future military technology and the West', *Royal United Services Institute Journal*, December, pp. 49–55.

28. Arnett, E. H. and Kokoski, R. (1993), 'Military technology and international security: the case of the USA', *SIPRI Yearbook*, pp. 307–29.

29. Bitzinger, R. A. (1994), 'The globalisation of the arms industry: the next proliferation challenge', *International Security*, **19** (2), pp. 170–98.

30. MccGwire, M. (1994), 'Is there a future for nuclear weapons?', *International Affairs*, **70** (2), pp. 211–28.

31. Anon. (1994), 'US exploring hi-tech non-lethal weapons', *The Times of India*, 2 February.

Chapter 3: The Inhumane Weapons Convention

1. Anon. (1994), 'Inhumane Weapons Convention', *Disarmament Newsletter*, **12** (3), pp. 11–12.

2. International Committee of the Red Cross (1994), *Report of the ICRC for the Review Conference of the 1980 United Nations Convention on Prohibitions or Restrictions on the use of Certain Conventional Weapons which may be deemed to be Excessively Injurious or to have Indiscriminate Effects*, ICRC, Geneva.

3. Pengelley, R. (1994), 'Wanted: A watch on non-lethal weapons', *International Defense Review*, April, p.1.

4. Speser, P. (1993), 'Public debate needed on laser weapons', *Laser Focus World*, September, p.1.

5. Kalshoven, F. (1987), *Constraints on the Waging of War*, International Committee of the Red Cross, Geneva.

6. Boyd, Lt. Col. P. M. (1991), 'The law of armed conflict; Definitions, sources, history', *Australian Defence Force Journal*, January/February, pp. 19–32.

7. Anon. (1993), *Victims of War*, ICRC, Geneva.

8. Bring, O. (1987), 'Regulating conventional weapons in the future – humanitarian law or arms control?', *Journal of Peace Research*, **24**, (3), pp. 275–86.

9. Goldblat, J. (1982), 'The laws of armed conflict: An overview of the restrictions and limitations on the methods and means of warfare', *Bulletin of Peace Proposals*, **13** (2), pp. 127–33.

10. Anon. (1973), 'The prohibition of inhumane and indiscriminate weapons', *SIPRI Yearbook* (Chapter 5), Almqvist & Wiksell, Stockholm.

11. Anon. (1975), 'The prohibition of inhumane and indiscriminate weapons', *SIPRI Yearbook* (Chapter 4), Almqvist & Wiksell, Stockholm.

12. Anon. (1979), 'The prohibition of inhumane and indiscriminate weapons', *SIPRI Yearbook* (Chapter 9), Almqvist & Wiksell, Stockholm.

13. Anon. (1980), 'The prohibition of inhumane and indiscriminate weapons', *SIPRI Yearbook* (Chapter 13), Taylor & Francis, London.

14. Röling, B. V. A. and Suković, O. (1976), *The Law of War and Dubious Weapons*, Almqvist & Wiksell, Stockholm, for SIPRI.

15. Lumsden, M. (1978), *Anti-personnel Weapons*, Taylor & Francis, London, for SIPRI.

16. Prokosch, E. (1995), *The Technology of Killing: A Military and Political History of Anti-personnel Weapons*, Zed Books, London.

Chapter 4: From Defensive Mines to Anti-Personnel Deserts

1. Doucet, I. (1993), 'The coward's war: landmines and civilians', *Medicine and War*, **9**, pp. 304–16.

2. Wurst, J. (1993), 'Ten million tragedies, one step at a time', *Bulletin of the Atomic Scientists*, July/August, pp. 14–21.

3. Gander, T. J. (1993), 'The mechanics of anti-personnel mines', in *Symposium on Anti-Personnel Mines* (Montreux, 21–23 April), ICRC, Geneva.

4. Anon. (1993), *Landmines: A Deadly Legacy*, Human Rights Watch Arms Project, New York.

5. Anon. (1995), *Making the World Unsafe for Landmines*, Project on Demilitarisation and Democracy, Washington, D.C.

6. Jefferson, P. (1993), 'Technical aspects of anti-personnel mines', in *Symposium on Anti-Personnel Mines* (Montreux, 21–23 April), ICRC, Geneva.

7. Office of International Security Operations (1993), *Hidden Killers: The Global Problem with Uncleared Landmines*, Bureau of Political-Military Affairs, Department of State, Washington, DC.

8. Macleod, S. (1993), 'And still they kill', *Time*, 13 December, pp. 24–30.

9. McGrath, R. (1991), *Report of the Afghanistan Mines Survey*, Mines Advisory Group, London.

10. Davis, P. and Dunlop, N. (1994), *War of the Mines: Cambodia, Landmines and the Impoverishment of a Nation*, Pluto Press, London.

11. Coupland, R. M. and Russback, R. (1994), 'Victims of anti-personnel mines: What is being done?', *Medicine and Global Survival*, **1** (1), pp. 18–22.

12. International Committee of the Red Cross (1993), *Symposium on Anti-Personnel Mines*, ICRC, Geneva.

13. British Red Cross (1995), *Focus: Anti-Personnel Mines*, January.

14. Dalton, D. (1994), 'Hidden killers', *Oxfam News*, Summer, p. 19.

15. Anon. (1994), *Anti-personnel land-mines: A scourge on children*, UNICEF, New York.

16. Boutros-Ghali, B. (1994),'The landmines crisis: A humanitarian disaster', *Foreign Affairs*, September/October, pp. 8–13.

17. Karniol, R. (1995), 'UN to legislate against anti-personnel mines', *Jane's Defence Weekly*, 11 February, p. 8.

18. Petracco, F. F. (1993), 'Anti-personnel mines, production and trading', in *Symposium on Anti-Personnel Mines*, ICRC, Geneva.

19. Rollo, Lt. Col. N. H. (1993), 'The military use of anti-personnel mines', in *Symposium on Anti-Personnel Mines*, ICRC, Geneva.

20. Prokosch, E. (1995), *The Technology of Killing*, Zed Books, London.

21. ICRC (1994), *Blinding Weapons: Gas 1918 . . . Lasers 1990s?*, ICRC, Geneva.

22. Anon. (1995), 'United States: U.S. Blinding Laser Weapons', *Human Rights Watch Arms Project* , **7** (5), May.

23. Anon. (1995), 'Programs and Markets: Chinese offer laser eye-damage weapon', *International Defense Review*, **5**, pp. 19–20.

Chapter 5: Lethal and Non-Lethal Chemical Agents

1. Adams, V. (1989), *Chemical Warfare, Chemical Disarmament: Beyond Gethsemane*, Macmillan, London.

2. Royal Society of Chemistry (1989), *Chemical Warfare*, Briefing Paper, London.

3. WHO Group of Consultants (1970), *Health Aspects of Chemical and Biological Weapons*, World Health Organisation, Geneva.

4. Maynard, R. L. (1995), 'Toxicology of chemical warfare agents', in B. Ballantyne *et al.* (eds.), *General and Applied Toxicology* (Vol. 2), Stockton Press, London.

5. Anon. (1991), *Medical Effects of Chemical Warfare Agents*, Working Party on Chemical and Biological Warfare, Cornwall, England.

6. Thompson, R. F. (1993), *The Brain: A Neuroscience Primer* (2nd edition), W. H. Freeman, New York.

7. Kruk, Z. L. and Pycock, C. J. (1991), *Neurotransmitters and Drugs*, Chapman & Hall, London.

8. Perry Robinson, J. (1971), *The Problem of Chemical and Biological Warfare: Volume I – The Rise of CB Weapons*, Almqvist & Wiksell, Stockholm (for SIPRI).

9. Military Agency for Standardisation (1973), *Medical Aspects of Nuclear Biological and Chemical Defensive Operations*, A Med P–6, NATO, Brussels.

10. SIPRI (1973), *The Problem of Chemical and Biological Warfare: Volume II – CB Weapons Today*, Almqvist & Wiksell, Stockholm (for SIPRI).

11. Harbour, F. V. (1990), *Chemical Arms Control: The US and the Geneva Protocol of 1925*, Case Studies in Ethics and International Affairs, No. 4, Carnegie Council on Ethics and International Affairs, New York.

12. Fuller, J. W. (1966), 'The application of international law to chemical and biological warfare', *Orbis*, **X** (1), pp. 247–73.

13. Mirimonoff-Chilikine, J. (1970), 'The Red Cross and biological and chemical weapons', *International Review of the Red Cross*, June, pp. 1–15.

14. Goldblat, J. (1970), 'Are tear gas and herbicides permitted weapons?', *Bulletin of the Atomic Scientists*, April, pp. 13–16.

15. Applegate, Col. R. (1969), *Riot Control – Matériel and Techniques*, Stackpole Books, Harrisburg, PA.

16. Security Planning Corporation (1972), *Nonlethal Weapons for Law Enforcement*, National Science Foundation, Washington DC.

17. Sweetman, S. (1987), *Report on the Attorney General's Conference on Less than Lethal Weapons*, National Institute of Justice, Washington DC.

18. Crichton, Wg. Cdr. D. *et al.* (1958), 'Agents for riot control: The selection of T. 792 (o-chloro-benzal malononitrile) as a candidate agent to replace CN', Porton Technical Paper No. 651, CBDE, Porton Down, UK.

19. Himsworth, H. *et al.* (1971), *Report of the Enquiry into the Medical and Toxicological Aspects of CS (Orthochlorobenzylidene Malononitrile). Part II: Enquiry into Toxicological Aspects of CS and its use for Civil Purposes*, HMSO, London.

20. Hu, H. *et al.* (1989), 'Tear gas – harassing agent or toxic chemical weapon?', *Journal of the American Medical Association*, 4 August, **262** (5), pp. 660-3.

21. Meselson, M. S. (1970), 'Chemical and biological weapons', *Scientific American*, May, pp. 304–13.

22. Verwey, W. D. (1977), *Riot Control Agents and Herbicides in War*, A. W. Sijthoff, Leyden.

23. Neilands, J. B. *et al.* (1972), *Harvest of Death: Chemical Warfare in Vietnam and Cambodia* (Appendix III), The Free Press, New York.

24. US Congress (1969), *Department of Defense Appropriations for 1970: Hearings before a Subcommittee of the Committee on Appropriations*, Part 6, House of Representatives, Ninety-First Congress, First Session, US Government Printing Office, Washington DC.

25. Boserup, A. (1973), *The Problem of Chemical and Biological Warfare: Volume III – CBW and the Law of War*, Almqvist & Wiksell, Stockholm (for SIPRI).

26. US Congress (1970), *Chemical – Biological Warfare: US Policies and International Effects*, Hearings before the Subcommittee on National Security Policy and Scientific Developments of the Committee on Foreign Affairs, House of Representatives, Ninety-First Congress, First Session, US Government Printing Office, Washington, DC.

27. Krause, J. and Mallory, C. K. (1992), *Chemical Weapons in Soviet Military Experience*, Westview Press, Boulder.

28. US Congress (1959), *Research in CBR (Chemical, Biological and Radiological Warfare)*, Committee on Science and Astronautics, House of Representatives, Eighty-Sixth Congress, First Session, US Government Printing Office, Washington DC.

29. Perry Robinson, J. *et al.* (1971), *The Problem of Chemical and Biological Warfare: Volume V – The Prevention of CBW*, Almqvist & Wiksell, Stockholm (for SIPRI).

30. US Army Chemical Corps (1960), *Summary of Major Events and Problems Fiscal Year 1959*, Historical Office, Army Chemical Center, Maryland.

31. Davis, S. L. (1970), *Riot Control Weapons for the Vietnam War*, Historical Monograph, AMC 56M, US Army Munitions Command, Edgewood Arsenal, Maryland.

32. Anon. (1988), *Israel and the Occupied Territories: The Misuse of Tear Gas by Israeli Army Personnel in the Israeli Occupied Territories*, Amnesty International (AI Index MDE/15/26/88), London.

33. Hogg, I. V. (1994), *Jane's Security and Co-in Equipment*, Jane's Information Group, London.

34. Fischetti, M. (1995), 'Less-than-lethal weapons', *Technology Review*, **14** January, pp. 14–15.

35. Wardle, C. (1994), 'Written Answers: "Pepper Gas"', *Hansard*, Col. 604, 15 June.

36. Conn, P. M. (1995), *Neuroscience in Medicine*, J. B. Lippincott Company, Philadelphia.

37. Bevan, S. J. *et al.* (1987), 'The mechanism of action of capsaicin – a sensory neurotoxin', in P. Jenner (Ed.), *Neurotoxins and their Pharmacological Implications*, Raven Press, New York.

38. Yanka, Col. D. E. (1960), 'Mickey Finn on the battlefield', *Army*, April, pp. 28–9.

39. Linsey, D. *et al.* (1960), '"Off the rocker" and "On the floor"', *Armed Forces Chemical Journal*, **14** (3), May–June, pp. 8–9.

40. Macy, R. (1956), 'The Chemical Corps search for chemical warfare agents', *Armed Forces Chemical Journal*, **10** (6), June, pp. 22–4.

41. Conahan, F. C. (1994), *Human Experimentation: An Overview on Cold War Era Programs*, Testimony before the Legislation and National Security

Subcommittee, Committee on Government Operations, House of Representatives, GAO/T–NSIAD–94–266.

42. Marks, J. (1980), *The Search for the 'Manchurian Candidate': The CIA and Mind Control*, McGraw-Hill, New York.

43. US Army Chemical Corps (1955), *Summary of Major Events and Problems Fiscal Year 1955*, Historical Office, Army Chemical Center, Maryland.

44. Office of the Assistant Secretary of Defense Research and Development (1955) *Report of the Ad Hoc Study Group on Psychochemical Agents*, PBC206/1, 19 November, Department of Defense, Washington DC.

45. US Army Chemical Corps (1956), *Summary of Major Events and Problems Fiscal Year 1956*, Historical Office, Army Chemical Center, Maryland.

46. US Army Chemical Corps (1957), *Summary of Major Events and Problems Fiscal Year 1957*, Historical Office, Army Chemical Center, Maryland.

47. US Army Chemical Corps (1959), *Summary of Major Events and Problems Fiscal Year 1958*, Historical Office, Army Chemical Center, Maryland.

48. US Army Chemical Corps (1962), *Summary of Major Events and Problems Fiscal Years 1961–62*, Historical Office, Army Chemical Center, Maryland.

49. Lee, M. A. and Shlain, B. (1985), *Acid Dreams: The CIA, LSD and the Sixties Rebellion*, Grove Press, New York.

50. Gerber, B. V. *et al.* (1964), *An Analysis of Potential Roles and Missions for Chemical Incapacitating Agents, Using CS, BZ, and 4X as Examples*, Chemical Research and Development Laboratories Special Publication 5–5, Edgewood Arsenal, Maryland.

51. Deseret Test Center (1972), *Joint CB Technical Data Source Book. Volume II: Riot Control and Incapacitating Agents. Part Three: Agent BZ*, DTC–TR–72–508, Fort Douglas, Utah.

52. Committee on Toxicology (1982), *Possible Long-Term Effects of Short-Term Exposure to Chemical Agents. Volume I: Anticholinesterase and Anticholinergics*, National Academy Press, Washington, DC.

53. Byrd, G. D. *et al.* (1992), 'Determination of 3–quinuclydinyl benzilate (QNB) and its major metabolites in urine by isotope dilution gas chromatography/mass spectrometry', *J. App.Toxicology*, **16** (3), pp. 182–7.

54. Rzeszotorski, W. J. *et al.* (1988), 'Affinity and selectivity of the optical isomers of 3–quinuclydinyl benzilate and related muscarinic antagonists', *J. Med. Chem.*, **31**, pp. 1463–6.

55. Hedges, C. (1995), 'Conflict in the Balkans: In Bosnia; Bosnia troops cite gassings at Zepa'. *The New York Times*, 27 July, Section A, p. 8, col. 4.

Chapter 6: The Human Nervous System

1. Thompson, R. F. (1993), *The Brain: A Neuroscience Primer* (2nd Edition), W. H. Freeman, New York.

2. Barker, R. A. (1991), *Neuroscience: An Illustrated Guide*, Ellis Horwood, Chichester.

3. Martini, F. H. (1995), *Fundamentals of Anatomy and Physiology* (3rd Edition), Prentice Hall, New Jersey.

4. Brody, T. M. *et al.* (1994), *Human Pharmacology: Molecular to Clinical* (2nd Edition), Mosby, St. Louis.

5. Deacon, T. W. (1992), 'Primate brains and senses', in S. Jones *et al.*, (Ed), *The Cambridge Encyclopedia of Human Evolution*, Cambridge University Press, Cambridge.

6. Deacon, T. W. (1992), 'The human brain', in S. Jones *et al.*, (Eds), *The Cambridge Encyclopedia of Human Evolution*, Cambridge University Press, Cambridge.

7. Bastian, G. F. (1993), *An Illustrated Review of the Nervous System*, Harper Collins, New York.

8. Fitzgerald, M. J. T. (1992), *Neuroanatomy: Basic and Clinical* (2nd Edition), Baillière Tindall, London.

9. Conn, P. M. (Ed.) (1995), *Neuroscience in Medicine*, J. B. Lippincott, Philadelphia.

10. Langston, J. W. (1986), 'MPTP-induced Parkinsonism: How good a model is it?', in S. Fahn *et al.* (Eds), *Recent Developments in Parkinson's Disease*, Raven Press, New York.

11. Gershon, E. S. and Rieder, R. O. (1992), 'Major disorders of mind and brain', *Scientific American*, September, pp. 127–33.

12. QED (1995), *A Hole in Fred's Head*, BBC1 TV, 21 August.

13. Barinaga, M. (1995), 'Missing Alzheimer's gene found', *Science*, **269**, 18 August, pp. 917–18.

14. Winkler, J. *et al.* (1995), 'Essential role of neocortical acetylcholine in spatial memory', *Nature*, **375**, 8 June, pp. 484–7.

15. Björklund, A. and Dunnett, S. B. (1995), 'Acetylcholine revisited', *Nature*, **375**, 8 June, p. 446.

16. Churchland, P. (1995), 'Perspective: Our three pound engine', *Times Higher Education Supplement*, 14 April, p. 16.

17. Perry Robinson, J. (1973), *The Problem of Chemical and Biological Warfare: Volume II – CB Weapons Today*, Almqvist & Wiksell, Stockholm (for SIPRI).

18. Woods, L. A. (1959), 'Incapacitation by anesthetic agents', in *Toxic Chemical Warfare Agents*, Report of Symposium IX, US Army Chemical Warfare Laboratories, CWL Special Publication No. 3, May (PB14 3503).

19. McNamara, B. (1959), 'Mechanisms of incapacitation', in *Toxic Chemical Warfare Agents*, Report of Symposium IX, US Army Chemical Warfare Laboratories, CWL Special Publication No. 3, May (PB14 3503).

20. Jaffe, J. H. and Martin, W. R. (1990), 'Opioid analgesics and antagonists', in

• REFERENCES •

L. S. Goodman and A. Gilman (Eds), *The Pharmacological Basis of Therapeutics* (8th Edition), Pergamon Press, New York.

21. Dando, M. R. (1994), *Biological Warfare in the 21st Century: Biotechnology and the Proliferation of Biological Weapons*, Brassey's, London.

22. Marshall, B. E. and Longnecker, D. E. (1990), 'General anesthetics', in L. S. Goodman and A. Gilman (Eds), *The Pharmacological Basis of Therapeutics* (8th Edition), Pergamon Press, New York.

23. Mestel, R. (1993), 'Cannabis: the brain's other supplier', *New Scientist*, 31 July, pp. 21–3.

Chapter 7: The Chemistry of the Brain

1. Brody, T. M. *et al.* (1994), *Human Pharmacology: Molecular to Clinical* (2nd Edition), Mosby, St. Louis.

2. Bloom, F. E. (1990), 'Drugs acting on the central nervous system', in L. S. Goodman and A. Gilman (Eds), *The Pharmacological Basis of Therapeutics* (8th Edition), Pergamon Press, New York.

3. Watson, S. and Girdlestone, D. (1994), *Receptor and Ion Channel Nomenclature Supplement (5th Edition): Trends in Pharmacological Sciences*, March, pp. 1–51.

4. Conn, P. M. (Ed.) (1995), *Neuroscience in Medicine*, J. B. Lippincott, Philadelphia.

5. Reisine, T. (1995), 'Review: Neurotransmitter receptors V: Opiate receptors', *Neuropharmacology*, **34** (5), pp. 463–72.

6. Peden, G. J. and Prys-Roberts, C. (1992), 'Editorial II: Dexmedetomidine – A powerful new adjunct to anaesthesia?', *British Journal of Anaesthesia*, **68**, pp. 123–5.

7. Thompson, R. F. (1993), *The Brain: A Neuroscience Primer* (2nd Edition), W. H. Freeman, New York.

8. Julien, R. M. (1995), *A Primer of Drug Action* (7th Edition), W. H. Freeman, New York.

9. Poling, A., Gadow, K. D. and Cleary, J. (1991), *Drug Therapy for Behavioral Disorders*, Pergamon Press, New York.

10. Feldman, S., Scurr, C. F. and Paton, W. (1993), *Mechanisms of Drugs in Anaesthesia* (2nd Edition), Edward Arnold, London.

11. Kruk, Z. L. and Pycock, C. J. (1991), *Neurotransmitters and Drugs* (3rd Edition), Chapman and Hall, London.

12. Franks, N. P. and Lieb, W. R. (1994), 'Molecular and cellular mechanisms of general anaesthesia', *Nature*, **367**, 17 February, pp. 607–14.

13. Woolverton, W. L. and Johnson, K. M. (1992), 'Neurobiology of cocaine abuse', *Trends in Pharmacological Sciences*, **13**, May, pp. 193–200.

14. Healy, D. (1993), *Psychiatric Drugs Explained*, Mosby Year Book, London.

15. Di Chiara, G. and North, R. A. (1992), 'Neurobiology of opiate abuse', *Trends in Pharmacological Sciences*, **13**, May, pp. 185–92.

16. Koob, G. F. (1992) 'Drugs of abuse: Anatomy, pharmacology and function of reward pathways', *Trends in Pharmacological Sciences*, **13**, May, pp. 177–84.

17. Baum, R. M. (1985), 'New variety of street drugs poses growing problem', *Chemical and Engineering News*, 9 September, pp. 7–16.

18. Emmett, J. C. (Ed.) (1990), *Comprehensive Medicinal Chemistry: Volume 3, Membranes and Receptors*, Pergamon Press, Oxford.

19. Concar, D. and Spinney, L. (1994), 'The highs and lows of prohibition', *New Scientist*, **1945**, 1 October, pp. 38–41.

20. Gelbard, R. S. (1994), 'Combating international narcotics trafficking', Statement to the House Foreign Affairs Committee, 22 June, *US Department of State Dispatch*, **5** (26), 27 June, pp. 440-4.

21. Stewart, D. P. (1990), 'Internationalizing the war on drugs: The UN Convention Against Illicit Traffic in Narcotic Drugs and Psychotropic Substances', *Denver Journal of International Law and Policy*, **18** (3), pp. 387–404.

22. Burger, A. S. (1995), 'Illicit narcotics and international crime: The effort to institutionalize controls', paper presented at the 36th Annual Convention of the International Studies Association, Chicago, February.

23. Flynn, Lt. Cdr. S. E. (1995), 'The erosion of sovereignty and the emerging global drug trade', paper presented at the 36th Annual Convention of the International Studies Association, Chicago, February.

Chapter 8: An Assault on the Brain

1. McCall, S. (1995), 'A higher form of killing', *Proceedings of the US Naval War College*, February, pp. 40–5.

2. Otter, T. (1992), 'Chemical warfare defence', *Military Technology*, 12, pp. 44–53.

3. Douglass, J. D. (1992), 'Who's holding the psychotoxins and DNA-altering compounds?', *Armed Forces Journal International*, September, pp. 50–2.

4. Lewis, D. A. *et al.* (1986), *Incapacitating Agents, European Communist Countries: Neurotoxins, Psychotropics, and Organofluorines*, Foreign Science and Technology Center, Army Intelligence Center, AST–1620R–100–86.

5. SIPRI (1982), 'The changing status of CB warfare', in *World Armaments and Disarmament: SIPRI Yearbook 1982*, pp. 317–62, Taylor and Francis, London.

6. SIPRI (1984), 'Chemical and biological warfare: Developments in 1983', in *World Armaments and Disarmament: SIPRI Yearbook 1984*, pp. 319–50, Taylor and Francis, London.

7. Contracts (1993), 'Less-than-lethal immobilizing chemicals', *ASA Newsletter*, **93–4**, p. 14.

8. Perry Robinson, J. P. (1994), 'Developments in "non-lethal weapons"

involving chemicals', in *Expert Meeting on Certain Weapon Systems and on Implementation Mechanisms in International Law*, ICRC, Geneva.

9. Perry Robinson, J. P. (1994), 'Disabling chemical weapons: Some technical and historical aspects', Second Workshop, Pugwash Study Group on Implementation of the CW Convention, 27–9 May, Netherlands.

10. Taylor, J R. and Johnson, W. N. (1976), *Use of Volunteers in Chemical Agent Research*, DAIG–IN 21–75, Office of the Inspector General and Auditor General, Department of the Army, Washington, DC.

11. Crossland, J. (1967), 'Psychotropic drugs and neurohumoral substances in the central nervous system', in G. P. Ellis and G. B. West (Eds), *Progress in Medicinal Chemistry* (Vol. 5), Butterworths, London.

12. McFarling, D. A. (1980), *LSD Follow-Up Study Report*, US Army Medical Department, US Army Health Services Command.

13. Committee on Toxicology (1982), *Possible Long-Term Health Effects of Short-Term Exposure to Chemical Agents*, National Academy Press, Washington, DC.

14. US Army Chemical Corps (1961), *Summary of Major Events and Problems*, Historical Office, Army Chemical Center, Maryland (quoted in reference 9).

15. Hudgens, G. A. and Holloway, W. R. (1971), *Behaviour Modification and Changes in Central Nervous System Biochemistry: An Annotated Bibliography*, Human Engineering Laboratories, US Army Research and Development Center, Aberdeen Proving Ground, Maryland.

16. Arthur D. Little Inc. and Sterling-Winthrop Research Institute (1967), *New Incapacitating Agents (Quarterly Report 15/16): Supplement 3. Preclinical Pharmacology and Toxicology of Candidate Agent 226,169*, Chemical Research Laboratory, US Army Edgewood Arsenal, Maryland (Contract No. DA18–108–AMC–103 [A]).

17. Pars, H. G. and Handrick, G. R. (1968), *New Incapacitating Agents: Final Summary Report – Part II. Index of Chemistry and Pharmacology of Compounds in Quarterly Reports 1–15–16, Volume I. Index by Subject, Compound Class and Compound Name. Volume 2. Index of Compounds by Number and Compound Numbers, Names, and Formulas (C–65401–FSR)*, Department of the Army, Edgewood Arsenal, Maryland (Contract No. DA18–108–AMC–103 [A]).

18. Jones, D. P. (1978), 'From military to civilian technology: The introduction of tear gas for civil riot control', *Technology and Culture*, **19** (2), pp. 151–69.

19. Whitten, B. (1968), *Nonlethal Agents in Crime and Riot Control*, Edgewood Arsenal Technical Memorandum EATM 133–1, Edgewood Arsenal, Maryland (quoted in reference 9).

20. Select Committee to Study Governmental Operations with Respect to Intelligence Activities (1975), *Unauthorised Storage of Toxic Agents* (pp. 12 and 192–7), US Senate, 16–18 September (quoted in reference 9).

21. Judiciary Subcommittee on Administrative Practice and Procedure and Senate

Labor and Public Welfare Subcommittee on Health (1975), *Biomedical and Behavioral Research, 1975: Human-Use Experimentation Programs of the Department of Defense and Central Intelligence Agency* (pp. 169–73, 1139, 1142), US Senate, 10 and 12 September and 7 November (quoted in reference 9).

22. Committee on Armed Services (1989), *National Defense Authorisation Act for Fiscal Years 1990 and 1991: Restoration of Certain Reporting Requirements in the Chemical and Biological Warfare Defense Programs* (p. 142), US Senate, 101st Congress, 1st Session, 19 July.

23. US Senate (1973), *Department of Defense Semiannual Report on Chemical Warfare and Biological Research Program Obligations for the Period Jan. 1 to June 30, 1973, Congressional Record*, 2 November, p. 9818.

24. Department of Defense (1990), *Annual Report on Chemical and Chemical/Biological Defense Research Program Obligations for the Period October 1, 1989 through September 30, 1990* (DD-USDRE (A) 1065), Washington, DC.

25. National Security and International Affairs Division (1990), *Chemical Warfare: DOD's Reporting of its Chemical and Biological Research*, GAO/NSIAD–90–102.

26. Presidential Statement to Congress (1979), *Fiscal Year 1980 Arms Control Impact Statements*, 96th Congress, 1st Session, March.

27. Yaniger, C. J. (1979), *Incapacitating Agent Weapons Technology, Final Report* (ARCSL-CR-79036), AAI Corporation, Baltimore, Maryland.

28. Monson, F. A. (1984), *Improved Design for a 155–mm Incapacitating Chemical Agent Munition, Final Report* (CRDC-CR-84126), AAI Corporation, Baltimore, Maryland.

29. Lydon, T. and Webb, M. (1986), *Chemical Warfare Defense Requirements: Worldwide Market Study and Forecast*, DMS (USA).

30. Scientific Advisory Board (1981), *Report of the Ad Hoc Committee on Potential New Methods of Detection and Identification of Chemical Warfare Agents* (Edgewood briefing, 'Characteristics of Chemical Agents', p. E4), United States Air Force, September (quoted in reference 9).

31. Department of the Army (1991), *Descriptive Summaries of the Research Development Test and Evaluation Army Appropriation: Supporting Data FY 1992/1993, Biennial Budget Estimate* (DA PAM 5–6–1), February.

32. Department of the Army (1992), *NBC Modernisation Plan*, Headquarters, Department of the Army, Washington, DC.

33. O'Malley, T. J. (1990), 'Ground warfare: Chemical warfare today', *Armada International*, **6**, pp. 20-7.

34. Wright, S. and Ketcham, S. (1990), 'The problem of interpreting the US biological defense research program', in S. Wright (Ed.), *Preventing a Biological Arms Race* (pp. 169–96), MIT Press, Cambridge, Mass.

35. Canada (1991), *Novel Toxins and Bioregulators: The Emerging Scientific and*

• REFERENCES •

Technological Issues Relating to Verification and the Biological and Toxin Weapons Convention, Canadian Government, Ottawa, September.

36. Factsheet (1995), *US Army Chemical and Biological Defense Command*, Public Affairs Office, Commander, CBDCOM, Edgewood Area, Aberdeen Proving Ground, Maryland.

37. Harford County Chamber of Commerce (1995), *Military Appreciation Week: Aberdeen Proving Ground*, 14–20 May.

38. Edgewood RDEC (1989–94), *Scientific Conference on Chemical and Biological Defense Research: Abstract Digest*, US Army Chemical and Biological Defense Command, Aberdeen Proving Ground, Maryland.

39. Mundy, B. P., Arnold, M. T. and Amend, J. R. (1993), *Organic and Biological Chemistry*, Harcourt, Brace, Jovanovich, New York.

40. Burger, A. (Ed.), (1970), *Medicinal Chemistry: Part II* (3rd Edition), Wiley-Science, New York.

41. Roth, A. J. and Kleeman, A. (1988), *Pharmaceutical Chemistry Vol. I: Drug Synthesis*, John Wiley, New York.

42. Bovill, J. G. (1991), 'Opioids', in J. W. Dundee, R. S. J. Clarke and W. McCaughey, (Eds), *Clinical Anaesthetic Pharmacology* (pp. 203–30), Churchill Livingstone, Edinburgh.

43. Vanio, O. (1989), 'Introduction to the clinical pharmacology of medetomidine', *Acta vet. scand.*, **85**, pp. 85–8.

44. Savola, J-M., and Virtanen, R. (1991), 'Central α_2-adrenoceptors are highly stereoselective for dexmedetomidine, the dextro enantiomer of medetomidine', *European Journal of Pharmacology*, **195**, pp. 193–9.

45. Bloor, B. C. *et al.* (1990), 'Sedative and hemodynamic effects of dexmedetomidine in humans', *Anesthesiology*, **73** (3A), pp. 411.

46. Mattila, M. J. *et al.* (1991), 'Effect of dexmedetomidine and midazolam on human performance and mood', *European Journal of Clinical Pharmacology.*, **41**, pp. 217–23.

47. Maze, M. and Tranquilli, W. (1991), 'Alpha-2 adrenoceptor agonists: Defining the role in clinical anesthesia', *Anesthesiology*, **74**, pp. 581–605.

48. Prolst, A. *et al.* (1984), 'Distribution of α_2-adrenergic receptors in the human brainstem: An autoradiograph study using [^{3}H]p-aminoclonidine', *European Journal of Pharmacology*, **106**, pp. 477–88.

49. Holets, V. R. (1990), 'The anatomy and function of noraderenaline in the mammalian brain', in D. J. Heal and C. A. Marsden (Eds), *The Pharmacology of Noradrenaline in the Central Nervous System*, Oxford University Press, Oxford.

50. Aston-Jones, G. *et al.* (1984), 'Anatomy and physiology of locus coeruleus neurons: Functional implications', in M. G. Ziegler and R. C. Lake (Eds), *Norepinephrine: Frontiers of Clinical Neuroscience, Volume 2*, Williams and Wilkins, Baltimore.

51. Correa-Sales, C. *et al.* (1992), 'A hypnotic response to dexmedetomidine, an α_2 agonist, is mediated in the locus coeruleus in rats', *Anesthesiology*, **76**, pp. 948–52.

52. Pertovaara, A. *et al.* (1994), 'Dissociation of the α_2-adrenergic antinociception from sedation following microinjection of medetomidine into the locus coeruleus in rats', *Pain*, **57**, pp. 207–15.

53. De Sarro, G. B. *et al.* (1987), 'Evidence that locus coeruleus is the site where clonidine and drugs acting at α_1- and α_2-adrenoceptors affect sleep and arousal mechanisms', *British Journal of Pharmacology*, **90**, pp. 675–88.

54. Rajbomski, J., Kubiak, P. and Aston-Jones, G. (1994), 'Locus coeruleus activity in monkey: Phasic and tonic changes are associated with altered vigilance', *Brain Research Bulletin*, **35** (5/6), pp. 607–16.

55. Air Force Office of Scientific Research (1995), *Research Interests and Broad Agency Announcement 95–1*, Bolling Air Force Base, Washington, DC.

56. Mohammad, F. K. *et al.* (1995), 'Reversal of medetomidine sedation in sheep by atipamezole and yohimbine', *Veterinary and Human Toxicology*, **37** (2), pp. 97–9.

57. Dose, V. A., Chen, B.-X. and Maze, M. (1989), 'Dexmedetomidine produces a hypnotic-anesthetic action in rats via activation of central alpha-2 adrenoceptors', *Anesthesiology*, **71**, pp. 75–9.

58. Meselson, M. (1992), *Banning Non-Lethal Chemical Incapacitants in the Chemical Weapons Convention*, Committee for National Security, Washington, DC.

59. Harvard-Sussex Programme (1995), *Chemical Weapons Convention Bulletin*, March 1996, p. 14 (entry dated 14–17 November).

Chapter 9: Arms Control for the 21st Century

1. Evans, G. (1994), 'Co-operative security and intra-state conflict', *Foreign Policy*, **96**, pp. 3–20.

2. Freedman, L. (1991), 'Arms control: Thirty years on', *Daedalus*, Winter, pp. 69–82.

3. Rogers, P. and Dando, M. R. (1990), *NBC 90: The Directory of Nuclear, Biological and Chemical Arms and Disarmament 1990*, Tri-Service Press, London.

4. Daalder, I. (1992), 'The future of arms control', *Survival*, Spring, pp. 51–73.

5. Roberts, B. (1992), 'Arms control and the end of the Cold War', *The Washington Quarterly*, Autumn, pp. 39–56.

6. Burns, R. D. (1992), *Encyclopedia of Arms Control and Disarmament*, Charles Scribner, New York.

7. Bertrand, M. (1991), 'The difficult transformation from "arms control" into a "world security system"', *International Social Science Journal*, **127**, pp. 87–102.

• **REFERENCES** •

8. Rogers, P. and Dando, M. R. (1992), *A Violent Peace: Global Security After the Cold War*, Brassey's, London.

9. Dando, M. R. (1994), *Biological Warfare in the 21st Century* (Chapter 5), Brassey's, London.

10. Dando, M. R. (1994), 'The challenge of new technologies', *The Indian Ocean Review*, **7** (2), pp. 1–7.

11. Bitzinger, R. A. (1994), 'The globalization of the arms industry: The next proliferation challenge', *International Security*, **19** (2), pp. 170-98.

12. Roberts, B. (1993), 'From non-proliferation to anti-proliferation', *International Security*, **18** (1), pp. 139–73.

13. Isaacs, J. (1995), 'Right says arms control wrong', *Bulletin of the Atomic Scientists*, September/October, pp. 18–19.

14. Isaacs, J. (1995), 'Cold warriors target arms control', *Arms Control Today*, September, pp. 3–7.

15. Lippman, T. W. (1995), 'Chemical arms pact in jeopardy', *Washington Post*, 13 November.

16. Metz, S. (1994), *The Revolution in Military Affairs and Conflict Short of War*, Strategic Studies Institute, US Army War College.

17. Metz, S. (1995), *Counterinsurgency: Strategy and the Phoenix of American Capability*, Strategic Studies Institute, US Army War College.

18. Lewer, N. (1995), 'Non-lethal weapons', *Medicine and War*, **11** (2), pp. 78–90.

19. Boyle, F. A. (1985), *World Politics and International Law*, Duke University Press, Durham, North Carolina.

20. Abbott, K. W. (1992), 'International Law and International Relations Theory: Building Bridges', *Proceedings of the 86th Annual Meeting*, pp. 167–72, The American Society of International Law.

21. Glaser, C. L. (1995), 'Realists as optimists: Cooperation as self-help' *International Security*, **19** (3), pp. 50-90.

22. Buzan, B. (1993), 'From international system to international society: Structural realism and regime theory meet the English school', *International Organization*, **47** (3), pp. 327–52.

23. Vander Lugt, R. D. (1995), 'International legal rules: Contending approaches', paper presented at the 36th Annual Convention of the International Studies Association, Chicago, February.

24. Beck, R. J. (1995), 'International law and international relations: The prospects for interdisciplinary collaboration', paper presented at the 36th Annual Convention of the International Studies Association, Chicago, February.

25. Weiler, L. D. (1986), 'General disarmament proposals', *Arms Control Today*, July/August, pp. 6–15.

26. Nadelmann, E. A. (1990), 'Global prohibition regimes: The evolution of

norms in intenational society', *International Organization*, **44** (4), pp. 479–526.

27. Dando, M. R. (1994), 'The management of international conflict', *American Behavioral Scientist*, **38** (1), pp. 133–53.

28. Harf, J. E. *et al.* (1995), 'Global population policy making after Cairo: An analysis of the population regime', paper presented at the 36th Annual Convention of the International Studies Association, Chicago, February.

29. Spector, L. S. and Dean, J. (1994), 'Cooperative security: Assessing the tools of the trade', in J. E. Nolan, (Ed.), *Global Engagement: Cooperation and Security in the 21st Century*, Brookings Institute, Washington, DC.

30. Perry Robinson, J. P. *et al.* (1993), 'The Chemical Weapons Convention: The success of chemical disarmament negotiations', *SIPRI Yearbook*, pp. 705–34, Oxford University Press (for SIPRI), Oxford.

31. Price, R. (1995), 'A genealogy of the chemical weapons taboo', *International Organization*, **49** (1), pp. 73–103.

32. SIPRI Yearbook (1993), 'Appendix 14A. The Convention on the Prohibition of the Development, Production, Stockpiling and Use of Chemical Weapons and their Destruction', pp. 735–56, Oxford University Press.

33. Dando, M. R. (1995), 'CBW Review: Problems and prospects in building an integrated arms control regime', *Brassey's Defence Yearbook 1995*, pp. 219–33, Brassey's, London.

34. Kellman, B. (1994), 'Bridling the international trade of catastrophic weaponry', *American University Law Review*, **43**, pp. 754–847.

35. Kamal, A. (1989), 'The Chemical Weapons Convention: Some particular concerns of developing countries', *Chemical Weapons Convention Bulletin*, **4**, pp. 1–2.

36. Rosenberg, B. H. (1994), '"Non-lethal" weapons may violate treaties', *The Bulletin of the Atomic Scientists*, September–October, pp. 44–5.

37. Editorial (1994), 'New technologies and the loophole in the Convention', *Chemical Weapons Convention Bulletin*, **23**, pp. 1–2.

38. News Chronology (1992) June 3, *Chemical Weapons Convention Bulletin*, **17**, p. 11.

39. News Chronology (1994) June 23, *Chemical Weapons Convention Bulletin*, **25**, p. 23.

40. News Chronology (1995) February 3, *Chemical Weapons Convention Bulletin*, **27**, p. 30.

41. Ramsbotham, D. (1994), *UN Peacekeeping: The Art of the Possible*, ISIS Briefing, **44**, ISIS, London.

42. Goldblat, J. (1987), 'The role of the United Nations in arms control: An assessment', in UNITAR, *The United Nations and the Maintenance of International Peace and Security*, pp. 369–85, Martinus Nijhoff, Lancaster.

43. Dahlitz, J. (1992), 'Legal issues concerning the feasibility of nuclear weapon

• REFERENCES •

elimination', in R. C. Karp, (Ed.), *Security Without Nuclear Weapons?*, Oxford University Press (for SIPRI), Oxford.

44. Schachter, O. (1994), 'United Nations Law', *The American Journal of International Law*, **88**, p. 1–23.

45. Törnudd, K. (1994), 'Integrating disarmament and arms regulation with UN peace activities', *Disarmament*, **XVII** (2), pp. 49–58.

46. Boutros-Ghali, B. (1994), 'Opportunities for global cooperation towards disarmament and arms control', *Disarmament*, **XVII** (2), pp. 1–10.

47. Carlsson, I. (1995), 'Roles for the UN in international security after the Cold War', *Security Dialogue*, **26** (1), pp. 7–18.

Chapter 10: Benign Interventions or a New Arms Race?

1. Capaccio, T. (1995), 'Non-lethal acquisition plan in the works', *Defense Week*, **16** (34), 21 August, pp. 1–9.

2. Anon. (1995), 'Perspectives', *Strategic Survey, 1994–1995*, Oxford University Press for IISS.

3. Binnendijk, H. and Clawson, P. (1995), 'New strategic priorities', *The Washington Quarterly*, **18** (2), pp. 109–26.

4. Owens, Admiral W. A. (1995), 'The emerging system of systems', *Proceedings of the US Naval War College*, May, pp. 36–9.

5. Parent, C. (1995), 'Draft policy directive spells out DOD's rules for non-lethal weapons', *Inside the Pentagon*, 13 July, pp. 1–2.

6. Jannery, B. (1995), 'No "silver bullet" solution in less-lethal weapons, Marine Corps say', *Inside the Navy*, 17 July, pp. 7–9.

7. Alexander, J. B. (1995), 'Non-lethal weapons and the future of war', paper presented at the Harvard–MIT seminar on *The Future of War*, Harvard University Center for International Affairs, 9 March.

8. Sherman, J. (1995), 'Army develops doctrinal framework for use of non-lethal capabilities', *Inside the Army*, **7** (29), 31 July, pp. 1–19.

9. Committee on Armed Services (1995), *National Defense Authorization Act for Fiscal Year 1996*, United States Senate, 104th Congress, 1st Session, 12 July.

10. Anon. (1995), 'ARPA wants proposals for developing non-lethal technologies', *Defense Daily*, 2 May.

11. Garwin, R. L. (1994), 'New applications of nonlethal and less lethal technology', in A. Kanters and L. F. Brooks, (Eds), *US Intervention Policy for the Post-Cold War World: New Challenges and New Responses*, W. W. Norton, New York.

12. Klaaren, Major J. W. and Mitchell, Major R. S. (1995), 'Nonlethal technology and airpower: A winning combination for strategic paralysis', *Airpower Journal* (Special Issue on Nonlethal Technology and Airpower), pp. 42–51.

13. Morris, C., Morris, J. and Baines, T. (1995), 'Weapons of mass protection:

Nonlethality, information warfare, and airpower in the age of chaos', *Airpower Journal*, Spring, pp. 15–29.

14. Nollinger, M. (1995), 'Surrender or we'll slime you', *Wired*, February, pp. 1–10.

15. Aftergood, S. (1995), 'A revolution in military affairs?', *F.A.S. Public Interest Report*, **48** (1), pp. 3–14.

16. Adelman, K. L. *et al.* (1995), *Non-Lethal Technologies: Military Options and Implications*, Report of an Independent Task Force, Council for Foreign Relations, Washington, DC.

17. Ness, L. (1995), 'Turf warfare scrutinized', *International Defense Review*, **5**, pp. 32–4.

18. Parent, C. (1995), 'Army pushes ahead with laser countermeasure system production', *Inside the Pentagon*, **11** (28), 13 July, pp. 9–11.

19. Evans, Representative L. (1995), 'Urge the Secretary of Defense to support an international ban on the use of lasers to intentionally blind as a method of warfare', Open Letter, US Congress, 11 July.

20. Hewish, M. (1995), 'Battlefield lasers: The race between action and countermeasure', *International Defense Review*, **2**, pp. 39–44.

21. Office of Assistant Secretary of Defense (Public Affairs) (1995), *DOD Announces Policy on Blinding Lasers*, (Reference No. 482–95), 1 September.

22. Goose, S. (1995), 'New US laser weapon policy: Welcome but inadequate', Open Letter, Human Rights Watch Arms Project, 6 September.

23. Parent, C. (1995), 'START II ratification on hold pending vote on State Dept. consolidation', *Inside the Pentagon*, **11** (34), 24 August, pp. 1–16.

24. Smit, W. A. (1991), 'Steering the process of military technological innovation', *Defence Analysis*, **7** (4), pp. 401–15.

25. Moodie, M. (1995), 'Beyond proliferation: The challenge of technology diffusion', *The Washington Quarterly*, **18** (2), pp. 183–202.

26. Barry, Col. J. L. *et al.* (1994), 'Nonlethal military means: New leverage for a new Era', National Security Program Policy Analysis Paper 94–01, J. F. Kennedy School of Government, Harvard University.

27. Dando, M. R. (1994), *Biological Warfare in the 21st Century*, Brassey's, London.

28. Cooper, P. (1995), 'Naval Research Lab. attempts to meld neurons and chips: Studies may produce army of "zombies"', *Defense News*, **10** (11), 20–26 March, pp. 1 and 50.

29. Collins, J. M. (1995), 'Nonlethal weapons and operations: Potential applications and practical limitations', US Congressional Research Service, 95–974S, 14 September.

30. Karsh, E. (1987), *The Iran-Iraq war: A Military Analysis,* Adelphi Papers **220**, IISS, London.

31. Hedges, C. (1995) 'Conflict in the Balkans: In Bosnia; Bosnia troops cite gassings at Zepa', *The New York Times*, 27 July, Section A, p. 8, col. 4.
32. Lederer, E. M. (1996), 'Journalist says Bosnian Serbs used banned gas in Srebrenica attack', *AP Datastream International News Wire*, 23 January. September.

INDEX